Library of
Davidson College

THE EVOLUTION OF THE ATOMIC THEORY

THE EVOLUTION OF THE ATOMIC THEORY

D. P. MELLOR

Emeritus Professor of Chemistry
University of New South Wales, Australia

ELSEVIER PUBLISHING COMPANY
Amsterdam London New York 1971

ELSEVIER PUBLISHING COMPANY
335 JAN VAN GALENSTRAAT, P.O. BOX 211, AMSTERDAM, THE NETHERLANDS

AMERICAN ELSEVIER PUBLISHING COMPANY, INC.
52 VANDERBILT AVENUE, NEW YORK, NEW YORK 10017

LIBRARY OF CONGRESS CARD NUMBER: 73-135496

STANDARD BOOK NUMBER: 0-444-40886-X

WITH 24 ILLUSTRATIONS AND 19 TABLES.

COPYRIGHT © 1971 BY ELSEVIER PUBLISHING COMPANY, AMSTERDAM

ALL RIGHTS RESERVED. NO PART OF THIS PUBLICATION MAY BE REPRODUCED, STORED IN A RETRIEVAL SYSTEM, OR TRANSMITTED IN ANY FORM OR BY ANY MEANS, ELECTRONIC, MECHANICAL, PHOTOCOPYING, RECORDING, OR OTHERWISE, WITHOUT THE PRIOR WRITTEN PERMISSION OF THE PUBLISHER,
ELSEVIER PUBLISHING COMPANY, JAN VAN GALENSTRAAT 335, AMSTERDAM

PRINTED IN THE NETHERLANDS

Preface

No scientist now doubts the existence of atoms. This statement could not have been made with the same degree of certainty even as recently as 1903, one hundred years after Dalton first put forward his chemical theory of atoms. During the present century, however, effects due to single atoms have been demonstrated in various convincing and sometimes dramatic ways; the nuclei of helium atoms leave their trails in the Wilson cloud chamber and photographic emulsions; the click of a Geiger counter announces the disintegration of a single radioactive atom; the scintillation of a crystal of zinc sulphide signals the impact of a single helium atom nucleus on the surface of the crystal.

Not only have the effects produced by single atoms been observed but direct images of individual atoms have been recorded in certain favorable instances. Images produced by means of the field-ion microscope have revealed arrays of close-packed atoms in the tip of an iridium crystal. A recent modification of this instrument, known as the atom-probe, field-ion microscope, enables an observer to select a single atom for study.

These striking achievements mark the latest chapters in a story that began more than two thousand years ago. This book attempts to sketch briefly the development of the atomic theory from its beginning in the fifth century B.C. up to the time when there was a more or less universal acceptance of the reality of atoms. As one might have anticipated, it proved difficult to locate this period with any degree of confidence. By the time some degree of unanimity about the existence of atoms had been reached, the Daltonian or "billiard ball" concept of the atom had been abandoned in favor of a divisible atom consisting of a positively charged nucleus surrounded by a counterbalancing cloud of negatively charged particles. The discoveries, mainly in the realm of physics, that brought about a revolution in our understanding of the nature and structure of atoms, at the same time did much to convert the sceptics to a belief in the existence of atoms. Since it would have fallen far outside the planned scope of this volume, no attempt has been made to trace even the early history of the growth in the understanding of atomic structure. Some of the discoveries that may have been responsible for the conversion of the sceptics are briefly referred to in the Epilogue.

A famous teacher of chemistry, Cannizzaro, once said "It often happens that the mind of a person learning a new science has to pass through all the phases which the science itself has exhibited in its historic evolution". This does not imply that the best way to study a subject is to begin with its history. But having once been introduced to the subject, something is to be gained by reinforcing its lessons with a study of its history.

In this way, it is hoped that the short account given here will prove useful to the more senior secondary school and university students of chemistry.

Acknowledgements

I wish to thank Professor A. N. Hambly and Dr. R. M. Gascoigne for their criticisms of drafts of the earlier chapters and Mr. D. Oldroyd for assistance in reading the proofs. The author alone is of course responsible for any shortcomings the book still possesses. My thanks are also due to Mr. V. Shuk for assistance with the diagrams and to Mrs. Marie Smith for typing the manuscript. Finally I am indebted to the Cambridge Univerity Press for permission to reproduce the many extracts from R. C. Trevelyan's translation of Lucretius' *De Rerum Natura* and to Dr. E. W. Mueller for permission to publish the ion-micrograph that appears on p. 162.

D. P. MELLOR

Contents

Preface v

Chapter 1. Origin of the atomic theory 1
Chapter 2. Revival of the Greek Speculations 14
Chapter 3. Prelude to the chemical theory of atoms 27
Chapter 4. Dalton and the chemical theory of atoms 44
Chapter 5. Avogrado partly overcomes a limitation of Dalton's theory 65
Chapter 6. The era of confusion and doubt 84
Chapter 7. The successful solution of the problem of determining atomic weights 104
Chapter 8. Consequences of solving the problem of atomic weights (I). Discovery of the periodic law 118
Chapter 9. Consequences of solving the problem of atomic weights (II). Birth of the structure theory 135
Chapter 10. Epilogue 156

Name index 165

Subject index 169

CHAPTER 1

Origin of the atomic theory

Historians generally agree that an atomic theory of matter originated with the Greeks. At this distance of time, it is difficult to be certain who first put forward the theory but the names of Leucippus, Democritus (born about 460 B. C.) and Epicurus who lived about 100 years later are usually prominently associated with its earliest development. Some authors believe that Leucippus, a shadowy figure historically, was the originator of the theory. Democritus, his successor, is better known. Since only a few fragments of the writings of Democritus have survived, his ideas which are thought to be an extension and an elaboration of those of Leucippus, are known through later writers. Of these, one of the best known is Epicurus of Samos (341–270 B. C.) who used the atomic theory as the basis of a materialist philosophy. The views of Epicurus have come down to us chiefly through the most extensive and valuable surviving source of information about the Greek atomic theory, namely, the poem *De Rerum Natura* (Concerning the Nature of Things) composed by the Roman poet Lucretius[1] about 57 B. C. *De Rerum Natura* has been variously described as "the greatest purely philosophical poem ever composed" and "an orderly treatise in physics". These descriptions hardly do justice to the poem's literary merit, but they do suggest a scientific interest which is unusual in poetry. In it one may discover the salient features of Greek atomic theory. Briefly summarised, they are:

atoms are exceedingly minute, indivisible, immutable particles which can neither be created nor destroyed; though built of some common substance, atoms differ in shape and size and weight; the space between atoms is empty (a vacuum); they collect into bodies by mechanical entanglement and attachment and so produce the infinite variety of the material world; the density of a body is in inverse relation to the amount of empty space between the atoms; atoms are in perpetual motion which persists by itself.

Before turning to the poem for an elaboration of these ideas, it is worthwhile to pause a moment to consider the question: "From what background of ideas did the Greeks develop the atomic theory?"

It is most unlikely that we shall ever know precisely what sparked off the flash of genius that led Leucippus to a theory, destined so long afterwards to exercise so profound an influence on science. There is, however, little doubt that the atomic theory had its origin in still earlier speculations about the nature of matter and that it was, in fact, one stage in attempts to resolve difficulties raised by these speculations.

For the beginning, we must go back to Thales of Miletus who lived in the 6th century B.C. He was the earliest member of a distinguished band of men who were the first to seek a rational explanation of the material world. One writer has said of him: "to all appearances he was the first human being who can be rightly called a man of science".

He seems to have had an intuitive belief that underlying the endless variety of the material world, there was some kind of unity. He set out to reconcile the "many" with "one" by seeking an answer to the question: Is there some primordial form of matter or primal element from which all other matter is derived? To which he gave the answer: "Yes—the inexhaustible substance from which everything is born and into which everything passes is water". Since none of the writings of Thales survives, we depend on later philosophers for hints as to the reasons why Thales chose water as the fundamental substance of the world.

Water is omnipresent—in the skies, on the earth and in the oceans. It is present in food, plants, animals and the soil, and plays a central part in the weather and the cycle of the seasons. It is essential for life. It is the only substance known to the ancients capable of existing in all three states of matter—gas, liquid and solid. For some or all of these reasons, Thales chose water. The importance of Thales lies rather in the nature of the question he asked than in the answer he gave to it. Scientists are still seeking an answer to his question.

Other later philosophers in their search for a "unifying principle" chose other substances; one believed that air was the entity from which all things were derived; another that everything was derived from varying combinations of earth, air, fire and water*. Then, returning

---

* For earth, air, fire and water, constituents thought to be present in all matter, the Greeks used the term elements. This notion of an element is, of course, quite different from the present day concept of a chemical element.

to the idea of "one", Anaximander suggested that all four were forms of a common indeterminate substance—a basic "universal stuff" that was entirely characterless.

Another fundamental problem considered by the Greek philosophers was the reconciliation of permanence and change in the material world. Parmenides asserted that there can be no such thing as change; whatever is, has existed and will exist for all time. Heraclitus, on the other hand, stated that all is change, that change or flux is a permanent condition of nature. Parmenides, seeking to discover some permanent substratum underlying all changes, was led to a view which was essentially this: that the amount of matter in the world is constant— that matter is never created from nothing nor is it ever completely annihilated. It was from this kind of thinking that the law of the conservation of matter ultimately emerged.

Leucippus and Democritus are thought to have arrived at a conceptual scheme, the atomic theory of matter, in an endeavour to resolve the dualisms of the earlier Greek thinkers: the relations of the One to the Many, of Permanence to Change. This they did by means of a new dualism: Atoms and Void (empty space)[2].

A fundamental question asked by the Greek atomists took the form of an imaginary or thought experiment. If one were to divide a piece of matter, a bar of metal for example, into smaller and smaller pieces, would there ultimately be a limit beyond which the process of subdivision could not be continued? About this question, there arose two opposing schools of thought: Leucippus and Democritus believed that matter possesses a granular or particulate structure and that there is a definite limit to the process of subdivision leading finally to ultimate particles or atoms*; the other school adopted the view sometimes referred to as the plenum theory, that matter possesses a perfectly continuous structure and that one could go on subdividing it forever— that is without limit. The supporters of the plenum theory of whom Aristotle (384—322 B.C.) was the most notable, gained the wider acceptance and held it for many centuries. It was not until the 17th century that the tide began to turn in favor of the atomic theory of the structure of matter.

In examining the arguments of the Greek atomists, one needs to be on guard against reading into their theory more than was, in fact,

---

* The word atom is derived from the Greek word atomos (indivisible).

there. Perhaps the best way of avoiding this danger and, at the same time, of gaining some insight into the way the Greeks thought about atoms is to study as nearly as possible exactly what they did say. This will be done by examining appropriate extracts from a translation of *De Rerum Natura* wherein Lucretius[1] expounds the views of Epicurus with probably some additions of his own.

Early in the poem Lucretius states the fundamental assumption of the theory:

> ... "the nature of the universe
> Consists then, in its essence, of two things;
> For there are atoms and there is void". (ref. 1, p. 16.) *

An essential notion here is that there must be empty space between atoms. Though the atomists were never able to demonstrate the existence of a vacuum and this failure proved a stumbling block to the acceptance of their views by other Greek philosophers, notably Aristotle, they were nevertheless able to bring forward arguments to suggest the existence of empty space between atoms.

> "Space then intangible, void
> And emptiness exist. If none existed,
> Then things would have no power to
> move at all.
> For that which is matter's function, to resist
> And hinder, would be present at all times
> To all things. Nothing could therefore advance
> Since no one thing would be the first to yield —
> But as it is we see before our eyes
> Through seas and land and through the heights
> of heaven
> That many things in many ways are moving". (ref. 1, p. 13).

Other arguments to demonstrate the existence of empty space between atoms were brought forward:

> "For without void nothing, it seems quite clear,
> Can be crushed in, or fractured, or by cutting,
> Be split in two". (ref. 1, p. 21).

If one uses a knife to cut an apple, the blade must find some empty space to penetrate the apple—so the argument went.

* The quotations are taken from R. C. Trevelyan's translation from the Latin.

The fundamental principle of Parmenides which provided one of the foundations of the atomic theory is succinctly stated:

> "And this first principle of her design
> Shall be our starting point; nothing is ever
> begotten by divine will out of nothing". (ref. 1, p. 7.)

> Furthermore nature dissolves each form back
> Into its own first bodies (atoms) nor does
> She ever annihilate things". (ref. 1, p. 9.)

While it is true that, in general, the Greeks did not perform experiments, they did observe nature closely. Most of their observations relating to the atomic theory seem to have been concerned with biological phenomena—birth, development and death. That matter is never completely destroyed would seem to be contrary to experience. Wood seems to disappear into nothing when it is consumed in a fire; plants and animals disappear on disintegrating after death. It is a tribute to the scientific intuition and imagination of the Greek atomists that they could see beyond appearances—that they were able to regard these seeming annihilations as deceptive.

It was Russell who once said (ref. 3 p. 63): "The search for something permanent is one of the deepest of the instincts leading men to philosophy." To explain the perpetual change in the world about them, the atomists saw indestructible and unchanging atoms as the permanent substratum of matter.

One argument ran as follows:

> "Again if time
> Utterly destroys, consuming all the substance
> Of whatsoever it removes from sight
> As the years lapse, out of what then does Venus
> Bring back into the light of life the race
> Of living creatures each after its kind?" (ref. 1, p. 9.)

> "But if
> Throughout that period of long time past
> Elements have existed out of which
> Our world of things is composed and remade
> Assuredly such atoms must be endowed
> With an immortal nature; none of them
> Therefore can turn to nothing". (ref. 1, p. 10.)

The compatibility of unity with variety of the material world was

References p. 13

achieved by the assumption that all atoms were built of the same "universal stuff" as envisaged earlier, by Anaximander. On this point, Lucretius makes very little, if any, comment.

Turning to another aspect of his exposition, we note that Lucretius was greatly impressed with the fact that like creatures produce like and in this repetition of forms of life, he found an argument in support of the theory.

> "Furthermore, since to each creature after its kind
> A limit of growing and maintaining life
> Has been ordained, and since by Nature's laws
> It stands decreed what they each can, and what
> They cannot do, and since nothing is changed,
> But all things are so far invariable
> That all the diverse birds one after another
> Show the peculiar markings of their kind
> Upon their bodies - they must certainly
> Also have bodies of unchangeable matter. (ref. 1, p. 23.)

Commenting on this passage, Vavilov[4] suggests that it may be interpreted thus: "If bodies consisted of an infinite number of infinitesimal or infinitely divisible parts, the number of possible combinations would be infinitely large and the repetition of previous forms would be very improbable." This would seem to be a powerful philosophical argument in support of the atomic theory. Today we are just beginning to understand the molecular mechanism that underlies heredity and the repetition of previous forms of life. However, some writers as far back as Cicero took just the opposite view—that the above observations provided evidence against the atomic theory. How, they argued, could a mere "fortuitous concourse of atoms" form themselves into a bird, for example; it was as though an infinite number of letters of the alphabet scattered about at random spontaneously formed themselves into *De Rerum Natura*. To accept Vavilov's interpretation without qualification may be to read more into Lucretius' words than is there.

If the process of disintegration is to be subsequently reversed, there must be a definite and finite limit to disintegration which is represented by atoms, otherwise the reverse process of birth and development could never be achieved. Only in this way would there remain building blocks for the generation of new life.

Thus, the disintegration of an organism after death was viewed as the destruction of grouping of atoms—birth as the reassembling into new but similar groupings of the same indestructible atoms, a view which has some resemblance to the modern one.

Concerning the sizes of atoms the Greeks had only the vaguest of ideas except that they seem to have agreed on one thing—that they were invisible. This is implied by Lucretius when he says "Nature operates by unseen bodies". Processes and changes take place, the reality of which cannot be doubted but the mechanism of which is beyond the powers of our senses to detect.

> "Then again
> We can perceive the various scents of things,
> Yet never see them coming to our nostrils
> ...
> Clothes also, hung up on a shore where waves
> Are breaking, become moist, and then grow dry
> If spread out in the sun. Yet in what way
> The water's moisture has soaked into them,
> Has not been seen, nor again in what way
> The heat has driven it out. The moisture therefore
> Is dispersed into tiny particles,
> Which our eyes have no power to see at all.
> Furthermore after many revolutions
> Of the sun's year, a finger-ring is thinned
> Upon the under side by being worn:
> Water falling in drops hollows a stone:
> The bent ploughshare of iron insensibly
> Grows smaller in the fields; and we behold
> The paving-stones of roads worn down at length
> By the footsteps of the people. Then again
> The brazen statues near the city gates
> Show right hands wearing thinner by the touch
> Of those who greet them ever as they pass by.
> Thus we perceive that all such things grow less,
> For they have been worn down: and yet what bodies
> Are leaving them each moment, this the grudging
> Nature of vision has precluded us
> From seeing. Finally whatever time
> And Nature add to things little by little,
> Obliging them to grow by slow degrees,
> No effort of our eyesight can behold.
> So too whenever things grow old by age
> Or through corruption, and wherever rocks
> That overhang the sea are gnawed away
> By the corroding brine, you cannot discern
> What they are losing at any single moment.
> Thus Nature operates by unseen bodies." (ref. 1, p. 12.)

References p. 13

The Greeks believed that atoms were in ceaseless motion.

> "For whenever
> As often happens in their swift career,
> They have met and collided, it results
> That at once they recoil in opposite ways:
> Nor is this strange; since they are very hard,
> Of heavy and solid mass, and from behind
> Nothing obstructs them . . ." (ref. 1, p. 46.)

While atoms themselves are too small to be seen, the observation of the effects of the incessant impacts on very small but visible bodies was, so the Greeks thought, possible. This is evident from a passage which is in many ways strongly reminiscent of a modern description of the Brownian movement. Lucretius (ref. 1, p. 63)* urges us to—"Observe what happens when sunbeams are admitted into a building and shed light on its shadowy places. You will see a multitude of tiny particles mingling in a multitude of ways in the empty space within the light of the beam, as though contending in everlasting conflict, rushing into battle rank upon rank with never a moment's pause in a rapid sequence of unions and disunions. From this, you may picture what it is for the atoms to be perpetually tossed about in the illimitable void. To some extent a small thing may afford an illustration and an imperfect image of great things. Besides, there is a further reason why you should give your mind to these particles that are seen dancing in a sunbeam: their dancing is an actual indication of underlying movements of matter that are hidden from our sight. There you will see many particles under the impact of invisible blows changing their course and driven back upon their tracks, this way and that, in all directions. You must understand that they all derive this restlessness from the atoms. It originates with the atoms, which move of themselves. Then those small compound bodies that are least removed from the impetus of the atoms are set in motion by the impact of their invisible blows and, in turn, cannon against slightly larger bodies. So the movement mounts up from the atoms and gradually emerges to the level of our senses, so that those bodies are in motions that we see in sunbeams, moved by blows that remain invisible."

* This prose quotation is from R. Latham's translation, Penguin Books, Harmondsworth, Middx., 1951.

The dancing of motes in a sunbeam is due to convection currents in the atmosphere and is not a manifestation of the Brownian movement. The irregular, ceaseless motion of particles under molecular bombardment can only be observed with particles too small to be seen with the naked eye but easily visible with a microscope. Only long after the invention of the microscope was this movement first observed, early in the nineteenth century. The English botanist Brown, in 1827, observed the incessant, random, hither and thither motion of pollen grains suspended in water, but was unable to account for it. It was not until towards the end of the nineteenth century that the motion was explained in terms of the kinetic theory of matter.

The atomists made no attempt to explain the origin of the motions of atoms. As Russell has pointed out (ref. 3, p. 85): "All causal explanations therefore must have an arbitrary beginning. This is why it is no defect in the theory of the atomists to have left the original movements of the atoms unaccounted for".

It has been said[5] that "the greatest achievement of the atomists was to develop a new kind of scientific reasoning based on evidence by analogy and inference of the invisible from the visible by means of parallels and models as illustrations." (See also ref. 6.) A good example of this is the picturesque description of the reason why an object, a bar of iron, for example, appears motionless although it consists of myriads of atoms in ceaseless motion.

> "And here you need not wonder why it is
> That though the primal particles of things
> Are all in motion, yet the whole sum appears
> To stand wholly at rest, except where something
> Is moving as one individual mass.
> For beyond our senses, far below their ken,
> Lies the whole nature of first elements.
> So since they are themselves invisible,
> Their motions too they needs must hide away;
> The more so that such things as we can see
> Will yet often conceal their motions from us,
> If they should be some distance away in space.
> Thus often on some hill the woolly flocks
> Creep onward cropping the glad pasturage
> Whichever way the grass pearled with fresh dew
> Tempts and invites each, while the full-fed lambs
> Gambol and butt playfully; yet they all
> Seem to us, blent by distance, to stand still
> Like a white patch upon the green hillside." (ref. 1, p. 54.)

References p. 13

All the wonderful variety of materials found in the world is ascribed to atoms of different sizes and shapes and their different groupings. Lucretius illustrates this by a striking analogy.

> "Nay in my many verses everywhere
> You see many letters common to many words;
> Yet you must needs admit that words and verses
> Are different both in meaning and in sound.
> So much can letters by mere change of order
> Accomplish; but those particles which are atoms
> Can effect more combinations, out of which
> All different kinds of things may be created." (ref. 1, p. 32.)

The analogy must not be pressed too far. In its modern form, different letters would correspond to atoms of different kinds of chemical elements and words would correspond to molecules. The Greeks had no idea of the modern concept of a chemical element and only the vaguest ideas of atoms clustering together to form finite groupings that we now call molecules. Not all combinations of atoms were possible. "You must not imagine" says Lucretius (ref. 1, p. 68), "that all atoms can be joined in all ways." It would, of course, be absurd to see in this statement the foreshadowing of the restrictions imposed on the combinations of atoms by valency rules of modern chemistry.

Having set down reasons for believing in the existence of atoms and void, the atomists attempted to describe various phenomena in terms of atoms endowed with various shapes and appendages, such as hooks or antler-like extensions by means of which they were attached to one another. Theirs was a mechanistic explanation of the world. They were the first to attempt to correlate physical properties of matter, for example hardness and density, with its atomic constitution.

Thus, hardness was explained in terms of the linking of atoms.

> "Furthermore things which seem to us hard and dense
> Must needs be made of particles more hooked
> One to another, and be held in union
> Welded throughout by branch-like elements.
> First in this class diamond stones, inured
> To despise blows, stand in the foremost rank,
> And stubborn blocks of basalt, and the strength
> Of hard iron, and brass bolts which, as they struggle
> Against their staples, utter a loud scream." (ref. 1, p. 59.)

Density is explained in terms of the closeness or otherwise of the packing of atoms.

> "Lastly, why do we see some things surpass
> Others in weight, though of no larger bulk?
> For if there be within a ball of wool
> Just as much substance as in a lump of lead,
> 'T is natural that both should weigh the same,
> Because the function of substance is to press
> Everything down, while on the contrary
> Void in its nature is always without weight.
> So when the one, though just as large, is found
> The lighter, this proves clearly that it holds
> More void; whereas the heavier of the two
> Shows that there is more substance in itself,
> And far less empty space." (ref. 1, p. 14.)

Water owes its fluidity, so the Greeks thought, to the fact that its atoms unlike those of diamond, are round and smooth and able to glide past one another.

> "But things that are liquid and of fluid substance
> Must consist rather of smooth round atoms;
> For the several globules do not hold together:
> And you may scoop up poppy seed as easily
> As water, which will also, if you spill it,
> Glide away with as ready a downward flow." (ref. 1, p. 59.)

The difference between the viscosity of wines and olive oil is explained:

> "And swiftly as we see wines flow through a strainer
> Yet sluggish olive oil will take its time,
> Undoubtedly because it is composed
> of atoms either larger or more hooked
> And mutually entangled." (ref. 1, p. 57.)

Apart from the attributes of size, shape and weight, atoms were thought to be "characterless". In attempting to explain the origin of color, the atomists rejected the idea of colored atoms; thinking of the white foam that capped blue ocean waves, Lucretius remarks:

> "But if the waters of the sea were made
> Of blue atoms there is no way whereby they
> could become white; for however much
> You jumble atoms together, atoms that are blue
> Never to a marble color can they change." (ref. 1, p. 71.)

If, he says, color depends on the shapes, positions and motions of atoms:

References p. 13

> "You can at once quite easily explain
> Why those things a little while before
> Were of a blue color, may suddenly
> Become of marble brightness." (ref. 1, p. 71.)

There can, of course, be no color without light

> "Nay a color changes
> In the light itself, according as it shines
> Struck by a straight or slanting beam of light
> So the plumage of doves, that ring their throats
> And crowns their necks, shows itself in the sun:
> Sometimes it is ruddy as lustrous garnet
> Then again, viewed some other way, it seems
> To mingle cobalt with green emerald." (ref. 1, p. 72.)

One of the principal aims of the Greek philosophers was to give a rational explanation of natural occurrences —"to explain them within the framework of general hypotheses." Undoubtedly, the greatest of these general hypotheses was the atomic theory, described by Russell (ref. 3, p. 84) as "more nearly that of modern science than any other theory propounded in antiquity."

It accounted for the phenomena of nature known at the time better than any other theory then current, in so far as it explained a large number of observations on the basis of relatively few assumptions.

Since there was, at the time of its origin, no deliberately designed, systematic experimentation of the kind that was to characterise the great scientific revolution of the seventeenth century, the atomic theory fell on barren ground. The seed thus sown was to lie dormant for many centuries and to germinate only after the introduction in the 18th and 19th centuries of the methods of systematic experimentation and quantitative studies which were to form the basis of modern chemistry.

It is perhaps unnecessary to add that the Greek atomic theory bears little resemblance to the modern theory. Throughout the nineteenth and the present century, it has undergone many changes: much has been discarded and those parts not discarded have been modified and refined. This is the fate to be expected for a theory so universal in its range of application and so fundamental in character. If it is reasonable to speak of the evolution of the atomic theory, then the Greek theory was the beginning or the first stage.

The Greek views about the structure of matter spread to the Arabs and the Hindus but it is not proposed to follow their fate in these countries, mainly because there appears to be no evidence that they were developed any further[7]. Christian and Jewish teachers often opposed atomism because of its association with atheism and materialism.

Suffice it to say that the atomic theory was never entirely forgotten though for many centuries interest in it fell to a very low ebb indeed. From time to time, in various parts of Europe, monks made copies of Lucretius' poem and these were preserved in monastic libraries. Two excellent 9th century manuscripts are still preserved. During the period 1000—1400 A.D., the existence of even these few manuscripts seems to have been largely forgotten.

REFERENCES

1 C. LUCRETIUS, *De Rerum Natura*, translated by R. C. TREVELYAN, Cambridge University Press, 1937.
2 L. L. WHYTE, *An Essay on Atomism*, Thomas Nelson and Son, London, 1951, p. 35.
3 BERTRAND RUSSELL, *The History of Western Philosophy*, George Allen and Unwin Ltd., London, 1961.
4 S. I. VAVILOV, *Newton Tercentenary Celebrations*, Cambridge University Press, 1947, p. 47.
5 S. SAMBURSKY, *The Physical World of the Greeks*. Translated from Hebrew by M. DAGINT, Routledge, London, 2nd edition, 1960, p. 128.
6 C. BAILEY, *The Greek Atomists and Epicurus*, Oxford University Press, 1931.
7 J. R. PARTINGTON, Origins of the Atomic Theory, *Ann. Sci.*, 4 (1939) 254.

CHAPTER 2

Revival of the Greek Speculations

Revival of interest in the atomic theory after a lapse of about fourteen centuries was part of the renewal of interest in Greek learning that took place during the Renaissance.

In the early years of the 15th century an Italian, Poggio Braccioloni, conducting a search for ancient manuscripts discovered a copy of *De Rerum Natura* in a German monastery. In 1414, he took the manuscript to Italy. Though this manuscript has since been lost, it was from it that the first printed edition of the poem was published in Brescia in 1473. It soon became well-known and between that year and 1600, there were no less than thirty editions; it was first translated into French in 1677 and into English in 1683.

It is evident that *De Rerum Natura* exerted a powerful influence on the thinking of the natural philosophers (scientists) of the 16th, and 18th centuries. Many notable men of the period—F. Bacon (1561–1626), P. Gassendi (1592–1655), T. Hariot (1560–1621), R. Boyle (1627–1691), W. Charleton (1619–1707) and I. Newton (1642–1727) to mention only a few, displayed a lively interest in the ideas of Democritus and Epicurus as expounded by Lucretius. Because of its far-reaching influence on the thinking of scientists, *De Rerum Natura* deserves a place among the great classics of science. There is little doubt that, mainly as a result of the influence of this book, the great majority of 17th century scientists accepted a particulate theory of matter. This theory was often referred to as the mechanical or corpuscular philosophy and the terms atom and corpuscle were often used interchangeably. Some writers, at least, drew a distinction between atoms and corpuscles, using the latter term to refer to small aggregations of atoms. It was Gassendi who first used the term molecule to describe a cluster of atoms. However, this usage should not be taken to correspond with its present day meaning, if for no other reason than that the concepts of chemical elements, and their characteristic atoms had not yet been developed.

Hariot was probably one of the first Englishmen to expound the atomic theory and, in doing so, he used it in an attempt to explain the refraction of light. Owing to the atheism of Epicurus and Lucretius, advocacy of the atomic theory was not without its dangers in the seventeenth century and its earliest protagonists were sometimes suspected of heresy. Indeed, it was because of this that Hariot was imprisoned in conditions that led to a serious deterioration of his health, and subsequently to his premature death which, according to one writer, was looked upon by some of his contemporaries as God's retribution for his impiety[1].

Gassendi, one of the foremost among the revivers of the atomic theory, avoided this difficulty by explaining that atoms had not been in existence from eternity but had been specially created by God. Newton also held this view (ref. 2, p. 400). "It seems probable to me, that God in the beginning formed matter in solid, massy, hard, impenetrable, moveable particles of such sizes and figures and with such other properties and in such proportion to space as most conduced to the end for which he formed them; and that these primitive particles being solids, are incomparably harder than any porous bodies compounded of them; even so very hard as never to wear or break in pieces; ordinary power being unable to divide what God himself made one in the first creation. While the particles continue entire, they may compose bodies of one and the same nature and texture in all ages . . . .

"But should the particles wear away or break in pieces, the nature of things depending on them would be changed. Water and earth composed of old worn particles and fragments of particles, would not be of the same nature and texture now, with water and earth composed of entire particles in the beginning. And therefore, that nature may be lasting, the changes of corporeal things are to be placed only in the various separations and new associations and motions of these permanent particles."

While Lucretius' poem *De Rerum Natura* did much to reawaken interest in the atomic theory, subsequent development of the theory was vitally influenced by the notable quickening of interest in experimental science at this time. This was the age of the great scientific revolution that heralded the birth of modern science.

References p. 26

Gassendi describes an interesting experiment in which he showed that water already saturated with salt could dissolve alum, thereby establishing, so he thought, the existence of pores or empty spaces between atoms. His experiment was criticised by one of his colleagues who showed that when alum is dissolved in saturated salt solution in a flask with a graduated neck, there is an increase in volume. Thus, the existence of pores was not proved. Apparently, no attempt was made to follow the experiment to a logical conclusion by comparing the volume of the added alum with the increase in volume of the solution.

One experiment that did much to focus attention on the atomic theory was Torricelli's invention of the mercury barometer and the consequent discovery of how to create a vacuum. This, it will be recalled, was a theoretical concept of the Greek atomists. Though at first there was a great deal of argument about whether the space above the mercury in a barometer was indeed empty, further experiments on the nature of this space eventually convinced many scientists that this was so. The invention of the air pump by von Guericke and subsequent construction of improved forms of the pump by Hooke and Boyle made available a second and more convenient method of creating a partial vacuum.*

In the 17th century there was probably no greater exponent of "experimental philosophy" than Boyle who, when speaking of his fellow scientists, said "that one of the most considerable services they could do the world is, to set themselves diligently to make experiments, and collect observations, without attempting to establish theories upon them before taking note of all the phenomena that are to be solved."

At a time when alchemy was not yet a spent force, Boyle did much to lay the foundations of modern chemistry. With his air pump, he was able to create a vacuum in a volume of space sufficiently large in which to carry out experiments. He showed that a flame burning in a large glass vessel was extinguished as air was removed from the vessel; that small animals and birds could not survive in an evacuated space and that the ticking of a clock was less audible in a vacuum than

---

* Even the best modern pumps cannot create a perfect vacuum. In this sense a perfect vacuum is still a mental abstraction.

in ordinary air. Though he did not discover phosphorus, he was the first man to investigate thoroughly and systematically its chemical and physical properties. He showed, for example, that it ceased to luminesce in the absence of air; also that luminous bacteria ceased to shine in vacuo.

Boyle has often been given the credit for being the first to define a chemical element but this is not justifiable. What he did was to challenge the ancient notion of the four elements—earth, air, fire and water —believed to be present in varying proportions in all things.

By means of numerous chemical experiments, described at length in his book *The Sceptical Chymist* Boyle[3] set out to show that this ancient view of the nature of matter could not be sustained. In this negative way, and positively through his foundational work on analytical chemistry which will be discussed in the next chapter, he helped to clear the way towards the understanding of some of the fundamental concepts of chemistry.

Our interest at this point, however, lies in another aspect of his work. One of Boyle's principal aims in his experimental work was to apply the corpuscular theory to chemistry.

Many scientists of the 17th century believed that heat is a mode of motion. Boyle, in particular, elaborated the idea that all the phenomena of nature could be explained in terms of matter and motion, insisting that motions of atoms and variations of their motions exercised an important influence on the properties of bodies. "Provided the minute parts (particles) be sufficiently agitated," he said "it matters not whether the motion be produced by fire or no: for by nimbly hammering iron or silver, you may put the minutest parts into such motion, as will make the metal very hot to touch." He believed that, in liquids, the atoms or corpuscles are in contact with one another but in continuous motion, whereas in solids they execute a vibratory motion*. This clearcut distinction between liquids and solids does not appear to have been made by the Greeks. Of the nature of the motion of particles in gases, Boyle was, curiously enough, not so certain. In 1661, he discovered the relationship between the pressure and volume of a gas (air) at constant temperature now referred to as Boyle's Law:

---

* These early ideas of heat as a mode of motion were later eclipsed for a time by another theory which, because it seemed to offer a better explanation of the facts, regarded heat as a material fluid — known as caloric.

References p. 26

The volume of a gas is inversely proportional to its pressure. He was not sure whether the spring of air (its pressure) was due to the fact that the particles of air behaved like little coiled springs or sponges or were widely separated and rapidly moving so exerting pressure by bombardment of a surface.

Boyle was, however, firmly convinced of the general importance of the motion of particles in explaining the phenomena of nature. For example, he explained the hygroscopic nature of certain salts in terms of the shapes and motions of particles without invoking the idea of attractive forces between them. A salt, calcium chloride, for example, absorbed moisture not because the particles of the salt attracted particles of water, a belief held later by Newton, but because the shape of the moving water particles enabled them to work their way into the pores (or inter-particle spaces) in the salt. Boyle applied a similar explanation to the process of solution. The particles of copper sulphate, for example, were similar in shape and size to the pores of the solvent water. Nitric acid dissolved silver, so Boyle thought, because the size and shape of the moving nitric acid particles permitted them to enter the pores present in the metal and so disintegrate its structure by some mechanism which he did not clearly delineate. Congruence between particles and pore size and shape did not exist with gold; therefore gold was not soluble in nitric acid—so the explanation ran.

Other chemists about this time gave explanations of chemical reactions in terms of particle shapes. Lemery,[4] the author of one of the first textbooks on chemistry believed that particles of acids possessed sharp spikes; particles of alkalies, on the other hand, were porous bodies into which the spikes of the acids penetrated to produce neutral salts. There was, of course, no experimental evidence whatsoever to support these highly imaginative views about the shapes of particles.

That the crude mechanical hook and eye picture of the attachment of one atom to another imagined by the Greeks in their attempt to explain cohesion made no appeal to Newton, the discoverer of gravitational attraction, is not surprising. He had explained the motions of planetary bodies in terms of gravitational attraction between them. Might not similar attractive forces act between atoms? Newton saw (ref. 2, p. 397) the possibility of providing a means of unifying all the phenomena of nature ranging from the heavenly bodies to the atoms of matter. "And thus nature will be very conformable to herself (*i.e.*

uniform) and very simple, performing all the great motions of the heavenly bodies by the attraction of gravity which intercedes those bodies and almost all the small ones of their Particles by some other attractive and repelling Powers which intercede the Particles". "There are" he said "agents in Nature able to make the particles of bodies stick together by very strong attractions and it is the business of experimental philosophy to find them out". On another occasion, he noted that "particles attract one another by some force, which in immediate contact is exceeding strong at small distances, performs the chymical operations (*i.e.* causes chemical reactions) and reaches not far from the particles with any sensible effect" (ref. 2, p. 389). These short-range forces appeared to be much stronger than gravity. The very rapid falling off with distance implies that the forces are not described by an inverse square law as is gravitational attraction. Newton seems to have suspected that they may be partly electrical in origin when he said "and perhaps electrical attraction may reach to such small distance, even without being excited by Friction." Additional evidence that the forces were not gravitational was to be found in their specificity; they were much stronger between some particles than between others which caused them to be referred to as "elective affinities".

The introduction in the 17th century of the concept of forces acting between atoms at a distance in a manner similar to the gravitational forces acting between heavenly bodies, was one of the first important modifications of the Greek atomic theory and one that was to occupy a permanent place in subsequent thinking about atoms. Its influence on scientists during the 17th and 18th centuries only will be considered here.

"How these attractions may be performed" said Newton (ref. 2, p. 376) "I do not here consider. What I call attraction may be performed by impulse or by some other means unknown to me. I use the word here to signify only in general any force by which bodies tend towards one another whatsoever be the cause. For we must learn from the phenomena of nature what bodies attract one another, and what are the laws and properties of attraction before we enquire the cause by which the attraction is formed."

Newton invoked the existence of inter-particle forces to describe both physical and chemical phenomena. The solubility of salt in water was attributed to attractive forces of different intensity; particles of salt were more strongly attracted to particles of water than they were to one another. To explain the formation of a crystal of salt during the evaporation of brine, Newton visualised (ref. 2, p. 388) the particles arranging themselves uniformly "in rank and file" as a consequence of their being acted upon "by some power which at equal distance is equal, at unequal distances, unequal." To explain the phenomenon of diffusion, he assumed (ref. 2, p. 387) the existence of repulsive forces between particles which came into play when the particles reached a certain distance apart and cohesive forces were no longer important.

"If a very small quantity of salt or vitriol (sulphuric acid) be dissolved in a great quantity of water, the particles of salt or vitriol will not sink to the bottom though they be heavier in specie than water but evenly diffuse themselves into all the water, so as to make it as saline at the top as at the bottom. And does not this imply that the parts of the salt or vitriol recede from one another and endeavour to expand themselves and get as far asunder as the quantity of water in which they float, will allow? And does not this endeavour imply that they have a repulsive force by which they fly from one another?"

Newton assumed the existence of repulsive forces between particles in his explanation of Boyle's Law. It will be recalled that Boyle himself favored the idea that the "spring of air" was due to the fact that the particles of air behaved like compressible coiled springs. On the other hand, both Hooke and Newton came near to the modern picture of a gas. In 1678, Hooke[5] attributed the pressure of a gas to the battering of its swift particles on the walls of its containing vessel, a view elaborated some sixty years later by Bernouilli[6]. Newton[7] believed "that the particles of vapours, exhalations and air, do stand at a distance from one another . . . separated from one another and kept at a distance by the said principle." The said principle was that the particles repel one another with a force varying inversely as the distance between them. On the basis of this assumption he demonstrated that the gas would obey Boyle's Law. He was not sure that the particles of a gas did repel one another in this manner: "Whether elastic fluids (gases) do really consist of particles so repelling one another is a physical question. We have here demostrated mathematically the properties of fluids consisting of particles of this kind . . .". Bernouilli[6], in 1738, using the assumption of widely separated but randomly moving particles gave a mathematical explanation of Boyle's Law very similar to the modern one. These were the first applications of mathematics to the atomic theory.

In the course of some experiments Bernouilli made the interesting discovery that air when subject to great pressure did not diminish in volume as much as would be excepted from the application of Boyle's Law. Lomonosov[8] pointed out that this was to be excepted if one took into account the fact that the particles of air possessed finite though small dimensions, thus anticipating by more than a hundred years a refinement of the kinetic theory of gases developed by van der Waals.

These views about the nature of the behaviour of particles in gases and solutions have been discussed at some length because they were destined to exercise a considerable influence on the thinking of John Dalton in the early part of the 19th century.

Evidence for preferential affinities or elective attractions, as they were sometimes called, was to be found not only in physical phenomena such as solution and diffusion but also in chemical reactions; for example, the replacement of copper by iron when a piece of iron is placed in copper sulphate solution. Newton suggested that the attraction or chemical affinity of the particles of iron for those of vitriolic (sulphuric) acid was greater than that of the particles of copper. Similar explanations applied for the remaining members, then known*, of what is now called the electrochemical series.

Another reaction which Newton explained (ref. 2, p. 378) in terms of elective attractions was the formation of spirit of salt (hydrogen chloride) by the treatment of salt with spirit of vitriol (sulphuric acid). "Does this not argue that the fixed alcaly of salt attracts the spirit of vitriol more strongly than its own spirit and not being able to hold them both, lets go its own." Newton gives many other examples but these are fairly typical.

All these reactions involved the motion of particles under the influence of attractive or repulsive forces. This is where Newton modified the ideas of Boyle. The latter explained the sharp, sour taste of acids by assuming, as did Lucretius, that the acid particles had sharp points that pricked the tongue. Newton, on the other hand, said that acids

---

* Newton lists (ref. 2, p. 381) iron, copper, silver, tin, lead and mercury. (Opticks 31st query.) The precipitation of silver from a solution of its nitrate by means of copper or mercury was known to the early alchemists.

References p. 26

had a sharp taste because under the influence of the force of attraction the spiky particles attained such speed as to bruise the tongue so producing the sharp sensation. In fact, he defined acids in terms of attractive force. "For whatever doth strongly attract, and is strongly attracted may be called an Acid."

In the 18th century the idea of elective attraction or relative affinity was further explored and generalised by Geoffroy (1672–1731). His basic notion was that "whenever two substances which have some disposition to unite, the one with the other, are united together and a third which has more affinity for the one of the two is added, the third will unite with one of these, separating it from the other". Applying this principle, Geoffroy drew up a table of affinities—the first of its kind. Still more elaborate tables were drawn up by other chemists, notably by T. Bergman (1735–1784), one of the great analytical chemists of the 18th century. Attempts were made by various chemists to give numerical values to affinities but so dubious was the experimental basis of these values that they provided no firm foundation for later investigations of any usefulness. The study of elective attractive forces between atoms and molecules proved to be an unfruitful line of attack at this stage in the evolution of the atomic theory with the result that towards the end of the 18th century, "the dream of chemistry as the quantified science of short range forces, the microcosmic counterpart of Newton's dazzling quantification of the macrocosm slowly faded away" (ref. 9).

In the latter half of the 17th century, the first attempts were made to estimate the upper limits of the sizes of particles in attenuated materials such as incense smoke, gold leaf and films of soap bubbles. Qualitative ideas about tenuous forms of matter go back as far as Lucretius. He pointed out that atoms were smaller than the motes in a sunbeam, smaller than the thickness of gold leaf or the thread of a spider web. More sophisticated qualitative ideas were developed by Boyle who noted that one part of copper in 513,620 volumes of water will, in the presence of ammonia, form an easily recognisable blue color. He also noticed that extremely small quantities of copper imparted a green color to a flame.

One of the first semi-quantitative estimates was made in 1664 by Charleton[10], physician to Charles II and author of one of the first books in English on the atomic theory. From a study of the smoke of frankincense, he estimated that there were more than 7×10^{17} particles in a grain of frankincense. Using results obtained in the gilding of silver wire, Halley[11] estimated that atoms of gold could not exceed $\frac{1}{34,500}$ of an inch and that a cube of gold $\frac{1}{100}$ inch on a side would contain 243,300,000 atoms of gold.

In the 1704 edition, Newton describes how he obtained an estimate of the upper limit of the size of soap particles based on the observations of the thickness of soap bubble film. His estimate of 10^{-5} cm is about

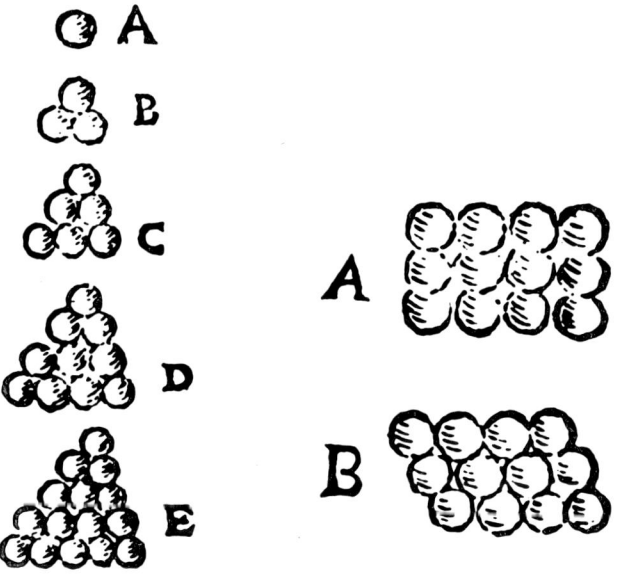

Fig. 1. From Kepler, "*A New Year's Gift of Hexagonal Snow*".

two orders of magnitude too high. "It is not impossible but that microscopes may at length be improved to the discovery of the corpuscles of bodies, on which their color depends. For if those instruments could be so far improved as with sufficient distinctness to represent objects five to six hundred times bigger than at a foot distance, they appear to our naked eye, I should hope, that we might be able to to discover some of the greatest of corpuscles". Many years were to elapse before microscopes of this power were developed. Indeed it was not until 1964 that the electron microscope was used, for example, to reveal the packing of the roughly spherical particles responsible for the brilliant colours of precious opal.

References p. 26

During the 17th and 18th centuries scientists speculated about the origin of the beauty of the form and symmetry of crystals.

The first man to publish the results of his speculations about the inner structure of a crystal was, curiously enough, the astronomer Kepler[12, 13]. In a short paper entitled *A New Year's Gift of Hexagonal Snow* and published in 1611, he imagined that the structure of a snow crystal as the result of the orderly close packing in space of equal spheres. The layers of particles sketched in his paper (Fig. 1) imply the idea of a crystal lattice though he did not express the idea. Kepler does not appear to have regarded the spherical particles as atoms in the Epicurean sense; nowhere in his paper did he use the term atom.

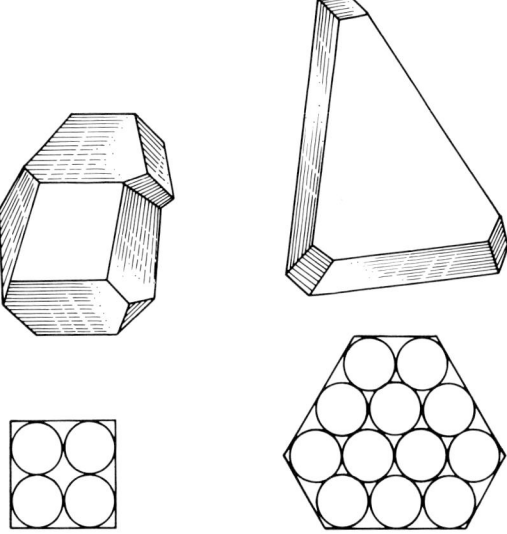

Fig. 2. A drawing of Hooke's representation of the structure of alum (based on *Micrographia*, pp. 85, 86).

Some 50 years later (1665) Hooke[5] revived and elaborated this idea extending it to other crystalline substances. He had the benefit of the microscope, newly invented, in observing crystals. In *Micrographia* where he gives one of the first accounts of the microscopic world he records his speculations about the structure of alum. He visualised crystals of alum as being built from equal, small, spherical particles which in two dimensional drawings he shows in regular close packing.

In a manner reminiscent of modern crystallography he attempted to relate the three dimensional structure of close-packed, equal spheres to the face development of the crystal (Fig. 2).

Like Kepler, Hooke does not seem to have regarded the spherical particles as atoms. He thought that salt (sodium chloride) and saltpeter (potassium nitrate) were similarly constituted of particles of the same size and shape as those of alum but he made no attempt to explain the origin of the chemical differences of these substances.

In an effort to explain the origin of the double refraction and cleavage of Iceland-Spar (calcite or calcium carbonate), Christian Huygens a few years later suggested that the building stones of this crystal were flattened spheroids or ellipsoids. These were also arranged in space in a regular, close-packed fashion. Neither the spheres nor the ellipsoids completely filled space even when in the closest packed array. This fact seems to have worried R. J. Haüy whom many regard as the father of modern crystallography. In 1784, Hauy postulated the existence, in crystals, of particles so shaped that when close-packed they completely filled space. When, for example, crystals of calcite are crushed, cleavage results in the formation of minute rhombohedra which can be packed together again completely filling space, reforming the original crystal. Haüy concluded that calcite was built from submicroscopic particles "constituent molecules" in the shape of rhombohedra. Confidence in this idea waned when it was found that some cubic crystals, for example calcium fluoride, showed octahedral cleavage—that is, cleavage fragments took the form of minute octahedra. Octahedra cannot be close-packed so as to completely fill space. None of these speculations about the structure of crystals, though correct in broad outline namely that their outer regularity and symmetry is a manifestation of an inner orderly arrangement of particles in space, proved fruitful in the further development of the atomic theory at that time. This was despite the fact that considerable advances were being made in understanding the external morphology and symmetry of crystals. The crystallographers of the day were unable to discover the nature and make up of the particles they wrote so much about.

Similarly the steps taken by Bernouilli[6] towards understanding the origin of the pressure exerted by gases on the walls of their containing vessels did not necessitate an understanding of the make up of their constituent particles. He was able to calculate the pressure of a gas in

References p. 26

terms of the number of particles in unit volume and the mass and velocity of each particle.

None of the avenues so far opened up—study of size, shape or interparticle forces was to prove profitable for the further immediate development of the atomic theory. The theory did however, possess what one writer has called a growing "expository value" meaning that it could be used to explain an increasing number of phenomena but it provided little or no opportunity for making testable predictions or incentives for further experimentation. Up to about the end of the 18th century, the atomic theory was mainly physical in character in the sense that it seems to have been more successful in explaining physical than chemical phenomena. This was so, for the simple reason that modern chemistry had not yet taken shape.

Much of the 17th and the greater part of the 18th century witnessed developments in chemistry which though not directed towards this end were, as can be seen in retrospect, essential for the emergence of a revolutionary form of the atomic theory, namely, the chemical theory of atoms. It will therefore be necessary to go back over the period already traversed to describe these events.

REFERENCES

1 H. R. KARGON, *Atomism in England from Hariot to Newton*, Clarendon Press, Oxford, 1966, p. 29.
2 I. NEWTON, *Opticks or a Treatise of the Reflections, Refractions, Inflections and Colours of Light*. Facsimile based on the 4th edition, London, 1730. Dover Publications Inc., 1952, Query 31.
3 R. BOYLE, *The Sceptical Chymist*, Everyman's Library Edition, J. M. Dent and Sons, London, 1949, p. 350.
4 N. LEMERY, *Cours de Chymie*, Paris, 1675.
5 R. HOOKE, *Micrographia*, Facsimile Edition, Weldon and Wesley Ltd. and Hafner Publishing Co., 1961, pp. 85–86.
6 D. BERNOUILLI, *Hydrodynamica*, Strasbourg, 1738, p. 200.
7 I. NEWTON, *Principia*, London, 1687, p. 301.
8 B. N. MENSHUTKIN, *Lomonosov*, Princeton University Press, Princeton, N. J., 1952
9 A. W. THACKRAY, Quantified Chemistry, The Newtonian Dream, a paper in D. S. L. CARDWELL, *John Dalton and the Progress of Science*, Manchester University Press, 1968, p. 93.
10 W. CHARLETON, *The Fabric of Science Natural, upon the Hypothesis of Atoms by Epicurus, Repaired by Petrus Gassendus, Augmented by W. Charleton*, Folio London, 1654; see also Ref. 1.
11 E. HALLEY, *Phil. Trans.*, 17 (1693) 540.
12 J. KEPLER, *A New Year's Gift of Hexagonal Snow*, 1611. German translation by H. STRUNZ AND H. BORN, Bosse, Regensberg, 1958.
13 C. SCHNEER, Kepler's Gift of a New Year's Snowflake, *Isis*, 51 (1960) 539.

CHAPTER 3

Prelude to the chemical theory of atoms

Looking back we can now see that an essential preliminary step toward the modern chemical theory of atoms, was the development of a number of fundamental chemical ideas: the concepts of pure substance, element, compound and chemical composition. These concepts emerged only very slowly and, except perhaps for the second, their formulation cannot be attributed to any individual chemist. It seems to have been more difficult for the human mind to grasp these concepts than those of force, mass, velocity and acceleration which were all well understood by about the middle of the 17th century. Another century or more was to elapse before the same could be said of the chemical concepts. The great revolution brought about by Lavoisier during which these fundamental chemical concepts were developed and effectively applied, occurred towards the end of the 18th century. It is the purpose of this chapter to describe very briefly some of the events leading up to, and as much of the revolution as seems relevant to the later growth of the atomic theory.

The term substance has a precise meaning in chemistry: it is any homogeneous species of matter of constant composition with characteristic physical and chemical properties which are invariable, being independent of its origin or mode of preparation*. Examples are gold, common salt (sodium chloride) and cane sugar (sucrose). An operational definition would describe a substance as the product of methods of separating, identifying and testing for purity**.

Chemistry has been defined[1] as "the science of substances—their structure, their properties and the reactions which change them into other substances". Thus, the concept of substance is essential to the

---

* When measured under constant conditions of temperature and pressure.
** An absolutely pure substance is an abstraction; in practice the limits of purity for ultra-refined metals which are among the purest substances known are in the neighbourhood of 99.999 %.

References p. 43

understanding of chemistry; without it and the derivative ideas of simple and compound substances, no chemical theory of atoms and molecules was possible.

Most of the material of our environment—the atmosphere, the oceans, the rocks, soil, plants and animals—consists of complicated mixtures of substances, many of which, especially those in plants and animals, can be separated from one another only with great difficulty. One of the basic and continuing tasks of chemistry is the isolation and identification of substances from these sources. Examples of relatively pure substances known to the ancient world were not very numerous comprising as they did some of the common minerals like quartz, and limestone, the metals gold, silver, copper, iron, mercury, tin and lead, and a few substances like salt, alum and soda (sodium carbonate) and, of course, water (as rain water). An important step towards grasping the idea of substances was their discovery in increasing numbers over the centuries and a growing acquaintance with their properties and transformations.

The long period during which the atomic theory was in eclipse (up to about the 15th century) saw the rise of alchemy which, strange though it may seem, has some relevance to the story of the atomic theory. Alchemy was dominated by the search for the Philosopher's Stone, an agent believed to be capable of transmuting base metals like copper, lead and mercury to gold.

Attempts to transmute base metals to gold are said to have had their origin in a number of observations: iron tools left in waters that accumulated in copper mines were found coated with copper. This gave rise to the idea that iron was being converted to copper. Copper, when alloyed with arsenic, resembles gold; this was taken to be a step towards the transmutation of copper to gold. A residue of silver or gold that sometimes remained on a prolonged heating of crude lead, as in the process of cupellation, was taken to indicate the generation of the precious metals from lead. The precious metals were, of course, originally present in the crude lead.

Futile as the search for the Philosopher's Stone proved to be, alchemy was not completely barren of results of significance for chemical practice. It is difficult to imagine a more powerful incentive to the study of the properties and transformations of matter than the promise of unlimited supplies of gold. Every imaginable kind of process was tried and though they all failed, some useful, if accidental, results were the outcome. The alchemists discovered many important substances; the mineral acids and a large number of their derivative salts, alcohol and phosphorus, arsenic, antimony and bismuth and their compounds, to name only a few.

Bacon[2] summed up the results of alchemy very neatly when he wrote: "Alchemy may be compared to the man who told his sons that he had left them gold buried somewhere in his vineyard; where they be digging found no gold but by turning up the mould about the roots of the vines procured a plentiful vintage. So the search and endeavours to make gold have brought many useful inventions and instructive experiments to light". (See also ref. 3.)

Perhaps the alchemist's greatest contribution to chemistry were laboratory techniques. The first laboratories were alchemical laboratories and in these alchemists developed the techniques of distillation, sublimation and especially crystallisation. By means of these techniques, they occasionally succeeded in separating mixtures into their component substances and thus they paved the way to the study of pure substances. "If we had to assess their position in the history of science," said Sherwood Taylor[3] when speaking of the alchemists, "we might best call them Fathers of Laboratory Technique".

It would be wrong to say that chemistry arose solely from alchemy. Modern chemistry is equally indebted to industries dependent on chemical operations, in other words, to industries like the extraction of metals, refining of precious metals, manufacture of glass and soap, the refining of sugar, salt and saltpeter (potassium nitrate), some of which were begun well before the Christian era. It is also indebted to the study of the use of chemicals for healing human ills, a subject once called iatrochemistry.

The origin of the idea of chemical composition—that a substance is composed of identifiable constituents—is not easily traced. The historian Guerlac[4] believes that "Glauber (1604–1670) was one of the first chemists, if not the first, to have a clear though only qualitative idea of chemical composition". Glauber recognized that ammonium chloride (sal ammoniac) was a compound of hydrogen chloride (spirit of salt) and ammonia (sal volatile urinae) and proved the point first by synthesising the salt and then decomposing it by heat into its constituents. About this time, it was also known that cinnabar (mer-

References p. 43

curic sulphide) could be formed by warming mercury and sulphur and that by strongly heating the resulting red powder, it could be decomposed to mercury and sulphur, each of which is readily identifiable.

An understanding of the nature of chemical composition was fostered by the growth of chemical analysis, which provided a reliable means of identifying substances. Qualitative analysis owes much to Boyle who advocated the use of a number of qualitative tests still in use today. He recognised ammonia by its smell and by the white fumes it produced when it came into contact with the vapours of hydrochloric acid; copper salts by the deep blue solution they gave with ammonia and the green color they imparted to a flame; salts of calcium and silver by the white precipitates they gave with sulphuric and hydrochloric acids respectively. He used the coloring matters of litmus, violets and cornflowers for the identification of acids; each of those substances, the forerunners of modern indicators, turned red in the presence of an acid. He identified some substances by their characteristic crystal forms.

Boyle measured the specific gravities of solids and liquids and so provided additional means, namely, the use of specific physical properties for assisting in the identification of substances*. The use of melting points for this purpose seems to have come later—probably towards the end of the 18th century.

An important objective of chemistry is the determination of the quantitative composition of substances. New substances are constantly being prepared in the laboratory or isolated from plants and animals. One of the first things a chemist does when studying a new substance is to determine its composition. It was not until about the middle of the 18th century that the more precise idea of quantitative composition was recognised. No one knows for certain which compound was first completely analysed quantitatively, with reasonable accuracy, but Guerlac[4] hazards a guess that it was cinnabar. In support of this suggestion, he quotes Macquer[5], who described cinnabar as being composed of seven parts by weight of mercury and one of sulphur.

---

* Archimedes was probably the first to use specific gravity as a means of identifying and establishing the purity of gold. The Arabs had also measured specific gravities.

Joseph Black (1728–1799) carried out what must have been among some of the earliest complete quantitative analyses of chemical substances. For example, he found that in 100 parts of soda crystals (sodium carbonate decahydrate), there are 20 of soda (Na_2O) and 16 parts of fixed air (CO_2) and 64 of water.

A. S. Marggraf (1709–1782) made an accurate determination of the weight of silver chloride precipitated when a known weight of silver was dissolved in nitric acid and the resulting solution treated with one of sodium chloride. In this case, however, Marggraf did not know all the constituents present in the precipitate; all he could say was it contained a certain percentage of silver. Experiments of this kind marked the beginning of quantitative analysis. The distinction between knowing the composition of a compound as a whole and the percentage of one particular constituent in a compound or a mixture is an important one. This latter had been carried out by assayers in the metal extraction industries for centuries. Nevertheless, the work of the assayer was important as establishing a tradition of accurate quantitative analysis. As Hall[6] put it: "Though more of a craftsman than a scientist and more concerned with utility than intellectual beauty, the assayer nevertheless collected a large part of the data on which modern chemical science was founded".

Boyle[7] has often been given the credit for being the first to define a chemical element. The following passage from his writings, usually proffered in support of this claim, is frequently quoted without the last portion which, of course, makes all the difference.

"And, to prevent mistakes, I must advertize [advise] you, that I now mean by Elements, as those Chymists that speak plainest do by their Principles, certain Primitive and Simple or perfectly unmingled bodies; which not being made of any other bodies, or of one another, are the Ingredients of which all those call'd perfectly mixed Bodies [*i.e.* compounds] are immediately compounded and into which they are ultimately resolved: now whether there be any such body to be constantly met with in all, and each, of those that are said to be Elemented bodies, is a thing I now question".

Boyle[7] devoted much of his book *The Sceptical Chymist* to casting doubt on the old idea of elements—earth, air, fire and water and later modifications of this idea, but he did not establish the modern concept. He never drew up a list of elements though at least nine were known

References p. 43

at the time. His thinking was along lines that excluded the modern idea of different chemical elements.

Boyle believed that the corpuscles of which all matter consisted were of the one stuff—that substances differed from one another in the shapes, arrangement and motion of corpuscles. The conclusion he drew from this was, when one comes to think of it, rather startling— but a consequence of the atomic theory of that day—"I see not why it should be absurd to think, that (at least among inanimate bodies) by the intervention of some very small addition or substraction of matter (which yet in most cases will scarce be needed) and of an orderly series of alterations, disposing by degrees the matter to be transmuted, almost of anything may at length be made of any thing". In other words by changing the shape, arrangement and motion of corpuscles, one could not only obtain gold from lead but anything from almost anything[8].

Boyle was familiar with the chemical behaviour of gold. Thus he wrote: "and the same gold will also by common aqua regia, and (I speak it knowingly) by divers other menstruums (*i.e.* solvents or reagents) be reduced into a seeming liquor (brought into solution) inasmuch as the corpuscles of gold will with those of the menstruum pass through cap paper (filter paper) and with them also coagulate into a crystalline salt (gold chloride)". He goes on to say: and in "many other ways gold may be disguised, and helped to constitute bodies of very different natures both from it and one another and nevertheless be afterwards reduced to the self-same numerical, yellow, fixt, ponderous and malleable gold before its commixture".

Boyle was also familiar with the characteristic, specific properties of gold by means of which it could be unfailingly identified and distinguished from all other substances: "If you ask a man what gold is and if he cannot show you a piece of gold, and tell you this is gold, he will describe it to you as a body that is extremely ponderous (dense), very malleable and fusible yet fixed in the fire (stable to heat) and of yellowish color; and if you offer to put off to him a piece of brass for a piece of gold, he will presently refuse it and (if he understands metals) tell you, that though your brass be colored like it, it is not so heavy or so malleable, neither will it like gold resist the utmost brunt of the fire, or resist aqua fortis (nitric acid)"[9].

Nevertheless the behaviour of gold was a great challenge to Boyle; in his experiments he continually sought ways of "debasing" gold, that is, of converting it to a base metal and *vice-versa* of transmuting base metals to gold.

Other great scientists of the day believed that it might be possible to effect the transmutation of metals. No less a person than Newton spent a great deal of time and effort in his laboratory at Trinity College, Cambridge, on this search[10]. At one stage of his work he so strongly suspected he had evidence of the transmutation of mercury to gold, that in 1689, he persuaded the Government to repeal an act of Henry VI banning efforts to make gold in the hope that if his suspicions were correct, the process would be legal. The result of his experience seems finally to have been that he thought transmutation theoretically possible but he was extremely sceptical about its practicability.

Throughout the centuries the nature of burning and flame was the source of much puzzlement and speculation. During the 18th century an attempt was made to explain it by what came to be known as the phlogiston theory. This theory, generally attributed to G. E. Stahl (1660–1734), had its origin in the study of metallurgical processes which were then largely centred in Germany. It was essentially a theory of combustion or oxidation. Metals were regarded as combinations of calx (the "ash" formed when a metal is strongly heated or burned in air) and a "principle" known as phlogiston. For example, when lead was heated in air, the result was a loss of phlogiston and the formation of a yellow powder.

lead → calx + phlogiston

When a mixture of the calx and finely divided carbon was heated, metallic lead reappeared.

The reaction between carbon which was thought to be rich in phlogiston, and calx was described thus:

calx + carbon → lead
    (source of
    phlogiston)

Phlogiston, believed to be present in all combustible materials, was regarded as an element (in the old sense of that word).

While a metal was regarded as a combination of its calx with phlogiston, in other words while a metal was regarded as more complicated than its calx, little progress could be made towards the modern concept of chemical elements.

References p. 43

The overthrow of the phlogiston theory resulted from three main developments:

(i) the application of the Newtonian idea that mass is a measure of the quantity of matter and the idea of weight;
(ii) the application of quantitative methods to the study of chemical phenomena—in other words, the quantification of chemistry;
(iii) the discovery of oxygen and the understanding of its role in the process of combustion.

Though he was by no means the first to carry out quantitative experiments in chemistry, Black's studies[11], about the year 1755, of the action of heat and acids on magnesia alba (basic magnesium carbonate) and limestone were a landmark in the history of chemistry providing as they did a model for later investigators. By means of a series of quantitative experiments, he showed clearly the relationships between and the nature of some of the important chemical reactions of magnesia alba ($MgCO_3Mg(OH)_2$) magnesia (MgO), limestone ($CaCO_3$), quicklime (CaO), slaked line ($Ca(OH)_2$), "fixed air" (CO_2), mild alkali (K_2CO_3) and caustic alkali (KOH)*.

One of his contemporaries said of Black[12], "I hesitate not to say that, excepting the optics of Newton, there is not a finer model for philosophical investigation". Black did, in fact, set the pattern for the quantitative studies later made by Lavoisier.

Black's work was important not only because of its exemplification of the quantitative method as applied to chemistry but also because it led to the discovery of carbon dioxide. Black was one of the first men to recognize a gas—"fixed air"—as a distinct chemical entity, different from common air**. The discovery of "fixed air", surprising as it may seem, was an outstanding advance in chemistry. The difficulty of taking this step was almost certainly due to the fact that the common gases are both colorless and odorless and therefore indistinguishable by means of the senses. Only by the discovery of some distinctive chemical test did this step become possible.

Black recognized "fixed air" as an individual substance by its ability to precipitate calcium carbonate from lime water, a reaction still used

* The chemical formulae in use today are given alongside the older trivial names to aid the reader. Such formulae came into use only at a much later date.
** Carbon dioxide was described early in the 17th century by Van Helmont who named it "gas sylvestre".

as a test for the gas. Since the quantification of chemistry and the discovery of the common gases and their chemical and physical properties were a prelude to the chemical atomic theory, it is worthwhile examining some of Black's work, though it is far too extensive to treat in detail.

Very briefly, he first studied the action of heat and acids on magnesia alba (basic magnesium carbonate) which he had prepared by treating a solution of Epsom salt ($MgSO_4$) with pearl ashes (K_2CO_3). He discovered that, on being strongly heated for some time, magnesia alba lost considerable weight being converted to magnesium oxide which on treatment with dilute sulphuric acid dissolved without the slightest trace of effervescence. In this respect, it differed from magnesia alba itself which effervesced violently when treated with dilute sulphuric acid though in both instances the solution so formed, yielded crystals of magnesium sulphate. The difference greatly puzzled Black for a long time.

Eventually he came to the conclusion that the result of heating basic magnesium carbonate was to expel "air" from it. It was for this reason, the resulting magnesium oxide no longer effervesced when treated with acid. The question then arose: whence came the "air"? He soon realised that it must have come originally from the potassium carbonate which also effervesced on treatment with acids. In one series of experiments, he took a known weight of basic magnesium carbonate, converted it to the oxide, thence to magnesium sulphate and finally by treating the last named with potassium carbonate obtained very close to the original amount of basic magnesium carbonate. In another series of experiments he showed that the loss of weight that occurred on mixing a known weight of limestone (calcium carbonate) with a weighed quantity of dilute hydrochloric acid was nearly identical with the loss of weight that took place when the same weight of limestone was strongly heated. In each instance the loss of weight was, of course, due to the escape of what he subsequently called "fixed air" (carbon dioxide).

For some time Black thought the effervescence and loss of weight on heating were due to the escape of ordinary air but gradually his suspicions were aroused that this was not so. He was aware that "fixed air" reacted with water saturated with lime turning it milky. He began to wonder whether the air ordinarily dissolved in pure water

References p. 43

was also absorbed by water when it was saturated with lime. To test this point, he carried out the following experiment. He placed two beakers, one containing pure water, the other clear lime-water (saturated solution of lime) in the receiver of a small air pump. As he proceeded to reduce the pressure in the receiver, air escaped from water and lime-water alike and, as far as he could judge, in about equal amounts. Thus, he concluded that lime-water liberated dissolved air but not the kind that reacted with the dissolved lime to form a milky white precipitate. "Quicklime therefore does not attract air when in its ordinary form but is capable of combining with one particular species only which is dispersed thro' the atmosphere, either in the shape of an exceedingly subtle powder or more probably in that of an elastic fluid (*i.e.* gas). To this I have given the name of "fixed air . . . . .". The name "fixed air", originally used by Hales, was intended to indicate the fact that a gas could be "fixed" or combined with a solid, in this case lime. That a gas could so enter the composition and become part of a solid made a great impression on some of Black's contemporaries, one of whom wrote[12]: "He (Black) has discovered a cubic inch of marble consisted of about half its weight of pure lime and as much "air" as would fill a vessel holding about six wine gallons . . . What could be more singular than to find so subtle a substance as "air" existing in the form of a hard stone and its presence accompanied by such a change in the properties of the stone?"

Other colorless, odorless gases were soon to be discovered and characterised by chemical and physical tests.

After Black discovered carbon dioxide, the isolation and characterization of other gases as distinct chemical entities, followed in quick succession. The most important were: hydrogen in 1766, by Cavendish (1731–1810); nitrogen in 1772, independently by Rutherford (1749–1819) and Scheele (1742–1786); oxygen in 1773–1774, independently by Priestley (1733–1804) and Scheele; chlorine in 1774 by Scheele.

The 18th century has often been referred to as the era of pneumatic chemistry for the reason that during this period, most of the common gases were isolated, characterized and recognized as individual substances—a process greatly assisted by the invention of a suitable method for collecting them. Modifying techniques used by Boyle and Mayow (1641–1761), Hales (1677–1761) devised the pneumatic trough in which gases were collected over water. Cavendish and Priestley later adapted this method for the collection of water-soluble gases by substituting mercury for water.

None of these gases was recognized as an element at the time of its discovery. The elementary nature of the first three was established by Lavoisier about 10 years later; that of chlorine about 36 years later (1810) by Davy.

Cavendish, Scheele and Priestley interpreted their discoveries in terms of the phlogiston theory.

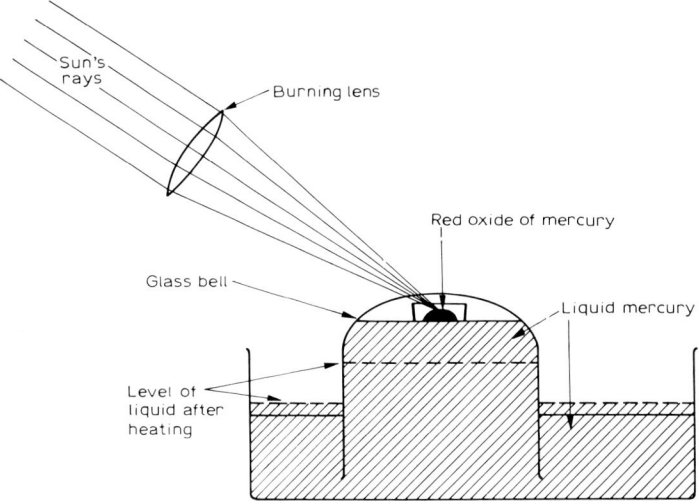

Fig. 3. The experimental arrangement used by Priestley for the isolation of oxygen (1774).

Hydrogen had been prepared by the action of acids on metals long before 1766, but Cavendish was the first to recognize it as a distinct chemical entity using mainly its inflammability and very low density as a means of characterizing it. He measured its density with considerable accuracy and showed that it had the same chemical properties irrespective of what acids and metals were used in its preparation. He named it "inflammable air" and believed it was pure phlogiston.

Priestley who isolated and described more gases than any other chemist made his great discovery in 1774 when he isolated oxygen by heating mercuric oxide in a confined space over mercury. The large burning glass which he used to heat the mercuric oxide by focussing

References p. 43

the sun's ray on it (see Fig. 3) is still to be seen in the Science Museum at South Kensington. Referring to this gas, he said[13] (see also ref. 14): "But what surprised me more than I can well express, was that a candle burned in this air with a remarkably brilliant flame." This property served to characterize the gas better than any other. Priestley called it "dephlogisticated air" and it is very doubtful whether he ever fully understood the significance of his discovery.

During a visit to Paris in October, 1774, Priestley described his discovery to Lavoisier who repeated the work and, subsequently, went on to make clear the role of oxygen in combustion. In order to indicate briefly how he did this, we must go back to some of Lavoisier's earlier experiments—more particularly to his work on the calcination of metals—their oxidation by heating in air. Other chemists had made a quantitative study of the calcination of metals many years before Lavoisier. Boyle, for example, heated metals in sealed glass vessels. He subsequently broke the seal and weighed the calx noting the increase in weight. He attributed the increase in weight to "fire particles" which had penetrated the glass. Lavoisier repeated these experiments with a simple but important variation in procedure. He carefully weighed a sealed glass retort containing small pieces of tin of known weight and then proceeded to heat it. He noted that after a while, calcination appeared to cease*. After cooling the retort, he again weighed it without breaking the seal. There was not the least change in weight. He broke the seal, noting at the same time a hissing sound from the inrush of air. He reweighed the retort and its contents and observed an increase in weight. He also weighed the metal and its calx and found that the increase in weight caused by the inrush of air was equal to the increase in weight of the metal and its oxide over that of the original metal. From the fact that there was no increase in weight of the sealed vessel, Lavoisier concluded that Boyle's interpretation of the experiment was not valid. Lavoisier had, in fact, demonstrated for the first time that, during a chemical reaction, matter is neither created nor destroyed but his attention was not fixed so much on this aspect of the experiment as on its bearing on the nature and composition of air.

His own view was that the increase in weight resulted from the

* The metal being present in excess was not completely converted to its calx but this did not affect the validity of the experiment.

"fixation" of part of the air by the metal, in other words, by the combination of the metal with part of the air. When Lavoisier heard Priestley's account of his work on dephlogisticated air (oxygen), he realized that the action of heat on *mercurius calcinatus per se*, (mercuric oxide prepared by heating mercury in contact with air), provided a means for examining the nature of that part of the air fixed during the calcination of a metal. This was not possible with the oxide of tin which unlike the oxide of mercury, cannot be decomposed by the sources of heat normally available in a laboratory.

Indeed, no other metal lends itself so readily to the study of the gas "fixed" during calcination as does mercury, and Lavoisier was quick to see this. It led him to carry out what was probably his most famous single experiment (see Fig. 4).

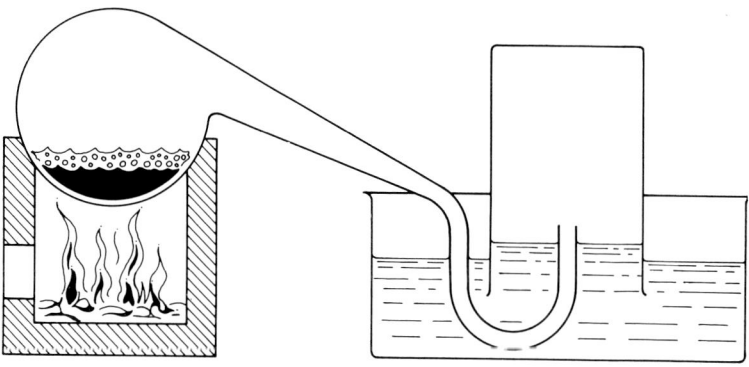

Fig. 4. The experimental arrangement used by Lavoisier for studying the composition of air.

He placed a weighed quantity of mercury in the retort and proceeded to heat it just below its boiling point for a period of 12 days. On allowing the whole apparatus to cool down to room temperature, he measured the volume of air that had disappeared owing to the formation of red mercuric oxide and found this to be approximately 8 cubic inches, about 1/6th of its original volume. He then took the mercuric oxide and heated it separately in an apparatus that enabled him to measure the volume of the resulting gas which had all the properties of Priestley's dephlogisticated air. The volume so measured was approximately 8 cubic inches. By means of this other experiments,

References p. 43

Lavoisier finally established: that air consists of at least two gases one of which (eminently respirable air—later called oxygen) combined with metals on calcination, thus causing an increase in weight; that this "air" was the active agent of combustion (and respiration); that metal calces were not elements but compounds of metals with oxygen. In so doing he clearly showed that the phlogiston theory was unsatisfactory and of no further use.

Another of Lavoisier's notable achievements was to establish the nature and composition of water. Cavendish was the first to show that water alone was produced by the burning of hydrogen. When, in 1783, after having cleared up the role of oxygen in the calcination of metals, Lavoisier learned of Cavendish's observation on the burning of hydrogen, he was able immediately to deduce the qualitative composition of water.

There has been a considerable amount of controversy about who should be credited with the discovery of the composition of water—a controversy complicated in part at least by the fact that Cavendish was a phlogistonist. The published facts are these: In 1784, Cavendish stated that inflammable air and dephlogisticated air combined to form water in volumes (measured under the same conditions of temperature) that stood in the ratio 201.5 : 100. Five years later, Lavoisier described an experiment in which he passed steam over a known weight of red hot iron. This enabled him to state the composition of water by weight as 85 parts of oxygen and 15 parts of hydrogen, a not very accurate result later used by Dalton.

Though he was the chief founder of modern chemistry, Lavoisier[15] was no great supporter of the atomic theory as may be seen from his discussion of elements. "I shall therefore content myself with saying that if, by elements, we mean to express the simple and indivisible molecules that compose bodies, it is probable that we know nothing about them: but if on the contrary, we express by the term elements or principles of bodies of the last point reached by analysis, all substances that we have not yet been able to decompose by any means are elements to us; not that we can assert that these bodies that we consider as simple are not themselves composed of two or an even greater number of principles, but, since these principles are not separated, or rather since we have no means of separating them, they are to us simple substances and we must not suppose them compounded until experiment

TABLE 1

LAVOISIER'S TABLE OF ELEMENTS

This table includes a number of substances (silica, magnesia, alumina and baryte) that are not now accepted as elements though, at that time, they met Lavoisier's criterion — they represented "the last point reached by analysis". Light and heat are of course not substances.

	Noms nouveaux.	Noms anciens correspondans.
Substances simples qui appartiennent aux trois règnes & qu'on peut regarder comme les élémens des corps.	Lumière.........	Lumière.
	Calorique.........	Chaleur. Principe de la chaleur. Fluide igné. Feu. Matière du feu & de la chaleur.
	Oxygène.........	Air déphlogistiqué. Air empiréal. Air vital. Base de l'air vital.
	Azote............	Gaz phlogistiqué. Mofete. Base de la mofete.
	Hydrogène.......	Gaz inflammable. Base du gaz inflammable.
Substances simples non métalliques oxidables & acidifiables.	Soufre...........	Soufre.
	Phosphore........	Phosphore.
	Carbone..........	Charbon pur.
	Radical muriatique.	Inconnu.
	Radical fluorique .	Inconnu.
	Radical boracique.	Inconnu.
Substances simples métalliques oxidables & acidifiables.	Antimoine........	Antimoine.
	Argent...........	Argent.
	Arsenic..........	Arsenic.
	Bismuth..........	Bismuth.
	Cobolt...........	Cobolt.
	Cuivre...........	Cuivre.
	Etain............	Etain.
	Fer..............	Fer.
	Manganèse.......	Manganèse.
	Mercure..........	Mercure.
	Molybdène........	Molybdène.
	Nickel...........	Nickel.
	Or...............	Or.
	Platine...........	Platine.
	Plomb............	Plomb.
	Tungstène........	Tungstene.
	Zinc.............	Zinc.
Substances simples salifiables terreuses.	Chaux............	Terre calcaire, chaux.
	Magnésie.........	Magnésie, base du sel d'Epsom.
	Baryte...........	Barote, terre pesante.
	Alumine..........	Argile, terre de l'alun, base de l'alun.
	Silice............	Terre siliceuse, terre vitrifiable.

and observation have proved them to be so". This last qualification was, in the event, well justified in the case of the oxides of magnesium, barium and calcium, all of which Lavoisier thought were elements. (See also ref. 16.)

To exemplify what he meant by an element, Lavoisier drew up the first list of chemical elements (see Table 1).

The table is taken from his textbook[17] *Traité Elémentaire de Chimie* which is a landmark in the history of chemistry.

In this book which stands in relation to chemistry much as Newton's *Principia* does to physics Lavoisier set out principles and ideas which provided the foundations of modern chemistry. He greatly clarified the presentation of these principles by adopting a rationalised chemical nomenclature which he had drawn up with the assistance of his colleagues, de Morveau, Berthollet (1748–1822) and Fourcroy. Lavoisier's great contribution to chemistry was that he explained quite simply the composition of chemical compounds, their reactions and the quantitative relations between them and gave them names that recalled their constituents.

In all his experimental work, Lavoisier assumed the truth of the law of the indestructibility of matter. Experiments like those on the calcination of metals were clear demonstrations of the truth of the law and on at least one occasion, he clearly enunciated the law-assertions to the contrary by some historians notwithstanding. "In every operation", he said, "there is an equal quantity of matter at the beginning and the end". He introduced the idea that a chemical reaction could be represented as a kind of balance sheet set out in the form of an equation which clearly implies the Law of the Conservation of Mass.

On one occasion, Lavoisier[18] complained of the slowness with which his antiphlogistic views were being accepted. He found consolation from the fact that their influence was greatest among the younger chemists: "I do not expect my ideas to be adopted all at once. The human mind gets creased into a way of seeing things. Those who have envisaged nature from a certain point of view during much of their career, rise only with difficulty to a new idea. It is the passage of time therefore which must confirm or destroy the opinions I have presented. Meanwhile, I observe with great satisfaction that young people are beginning to study the science without prejudice".

REFERENCES

1. L. Pauling, *General Chemistry*, W. H. Freeman and Co., San Francisco, 1956, p. 1.
2. F. Bacon, *De Augmentis Scientiarum. The Philosophical Works of Francis Bacon of the Advancement of Learning*. George Routledge and Sons Ltd., London, 1905, p. 57.
3. F. Sherwood Taylor, *The Alchemists*, Henry Schuman, New York, 1940, p. 190.
4. H. Guerlac, Quantification of Chemistry, *Isis*, 52 (1961) 194.
5. P. J. Macquer, *Dictionnaire de Chymie*, 1766.
6. A. R. Hall, *The Scientific Revolution* 1500–1800, Longmans, London, 1962, p. 223.
7. R. Boyle, *The Sceptical Chymist*, Everyman's Library Edition, J. M. Dent and Sons, London, 1911, p. 187.
8. T. S. Kuhn, Robert Boyle and Structural Chemistry, *Isis*, 43 (1952) 22.
9. R. Boyle, Origin of Form and Qualities, *Works*, Vol. II, p. 487.
10. R. J. Forbes, Was Newton an Alchemist?, *Chymia*, 2 (1949) 27.
11. J. Black, *Experiments on Magnesia Alba*, Alembic Club Reprint, No. 1, Livingstone Ltd., Edinburgh, 1944, pp. 1–46.
12. J. Read, *Joseph Black, the Teacher and the Man*, in A. Kent (Ed.), *An Eighteenth Century Lectureship in Chemistry*, Glasgow University Publications, Jackson and Sons, Glasgow, 1950, p. 81.
13. J. Priestley, *The Discovery of Oxygen*, Alembic Club Reprint, No. 7, Livingstone Ltd., Edinburgh, 1947, p. 10.
14. F. W. Gibbs, *Joseph Priestley, Adventurer in Science and Champion of Truth*, Nelson, London, 1965.
15. A. Lavoisier, *Elements of Chemistry*, translated by Robert Kerr, 1790. Reprinted in facsimile by Dover Publications Inc., New York, 1965, pp. 17–18.
16. D. McKie, *Antoine Lavoisier, Father of Modern Chemistry*, Lippincott, Philadelphia, 1935.
17. A. Lavoisier, *Traité Elémentaire de Chimie*, Paris, 1789.
18. A. Lavoisier, Reflections on Phlogiston, Serving to Develop the Theory of Combustion and Calcination, *Mém. Acad. Roy. Sci.*, (1783) 505.

CHAPTER 4

Dalton and the chemical theory of atoms

One of the most notable of Lavoisier's contemporary young supporters was the Irish chemist William Higgins (1763–1825) who, at the age of 26 published a book[1] entitled *A Comparative View of Phlogistic and Anti-Phlogistic Theories*. The year of its publication coincided with that of the publication of Lavoisier's *Traité Elémentaire de Chimie*. However, the interest of Higgins' book lies not so much in his defence of Lavoisier's views as in his (Higgins') attempt to apply atomic theory to chemical problems. There is no doubt that Higgins was a highly original thinker and that, by making use of many of Lavoisier's ideas concerning elements and compounds, he applied the atomic theory to chemistry with greater insight than anyone before him. Some of his innovations may, at first sight, appear trivial. They were, in fact, not only original but essential to the development of a chemical theory of atoms. He was the first to interpret elements and compounds in terms of atoms and molecules. He assumed that all atoms of a given element are identical: that atoms of different elements are different from one another, though not necessarily different in weight. To Higgins belongs the credit also of being the first to use the initial letters*, with or without other letters in the name of an element, to represent an atom of an element in a molecular formula. His symbols differ from those in present day use; for example, he used I for iron, C for copper, M for mercury, D for oxygen (dephlogisticated air) and so on. He thought of compound formation as the result of an atom to atom linking and used lines to represent the forces (in modern terms, chemical bonds) linking atoms together in molecules. His molecular formula for copper oxide was C–D; for sulphur dioxide S–D and sulphur trioxide $S\genfrac{}{}{0pt}{}{\diagup D}{\diagdown d}$. Where he believed there were two atoms of the one kind in a molecule,

---

* Letters had been used to represent elements even earlier. Hasenfratz and Adet used letters with circles, *e.g.* Fe and Sb to represent iron and antimony but these symbols did not represent atoms of an element.

he distinguished between them by using small and capital letters, respectively (ref. 2, pp. 88, 89).

Like other atomists* of the 18th century, he concentrated a great deal of his attention on the attractive forces between atoms. Several years earlier, his uncle Bryan Higgins had stated his belief that the attractive forces between atoms varied inversely as the nth power of the distance between them where n presumably was greater than 2. Higgins represented the inter-atomic forces linking one atom to several others, as being divided in a curious and arbitrary manner and, in a vague kind of way, he seems to have foreshadowed the idea of valency bonds. We have already seen that study of inter-atomic forces at this point of time proved unprofitable. It was not until early in the 20th century that chemists began to understand the nature of the forces that attracted one atom to another and held them together to form molecules.

Some writers have suggested that Higgins' formulae for sulphur dioxide and sulphur trioxide and for the various oxides of nitrogen foreshadowed the law of multiple proportions. However, he never explicitly stated this law.

In view of his later claims to have anticipated Dalton in originating the chemical atomic theory it is worth examining in a little detail Higgins' views about the all important topic of the weights of atoms. A good example of these views is provided by his discussion of the composition of sulphur dioxide (ref. 2, p. 72). Therefore, he said "100 grains of sulphur, making an allowance for water, require 100 or 102 grains of the real gravitating matter of dephlogisticated air (oxygen) to form volatile vitriolic acid (sulphur dioxide); and as volatile vitriolic acid is very little short of double the specific gravity of dephlogisticated air, we may conclude, that the *ultimate particles of sulphur and dephlogisticated air contain equal quantities of solid matter*; for dephlogisticated air suffers no considerable contraction by uniting to sulphur in the proportion merely necessary for the formation of volatile vitriolic acid. Hence, we may conclude that, in volatile vitriolic acid, *a single ultimate particle of sulphur is intimately united to a single particle of dephlogisticated air*".

---

* Among these, R. J. Boscovich (1711–1787) a Jesuit mathematician and astronomer, was the most notable. He assumed that atoms were mere mathematical points-centers of force and presumably without mass. His theory made no quantitative predictions and probably had more influence on physics than chemistry.

Higgins' experimental data are very close to those accepted today. Oxygen does indeed combine with an equal weight of sulphur to form its own volume of sulphur dioxide. An interesting point in Higgins' discussion is his use of a comparison of gas densities to draw conclusions about the relative weights of ultimate particles. The details of the reasoning which led him to the conclusion that one particle of sulphur combines with one particle of oxygen to form sulphur dioxide are not clear. Partington (ref. 2, p. 72) suggests that a possible interpretation of the passage quoted is that Higgins "had a vague idea that equal volumes of gases under the same external conditions contain in some instances the same number of particles" but Higgins himself never said so explicitly.

Higgins' formula for sulphur dioxide, expressed in modern symbols, is SO. It so happens that an atom of oxygen is almost exactly half .he weight of an atom of sulphur and that in sulphur dioxide, sulphur is linked to two atoms of oxygen. It is therefore, easy to see why Higgins believed that the atoms of oxygen and sulphur are equal in weight. Whether he believed that the atoms of all elements have the same weight, as Roscoe and Schorlemmer have suggested, is another matter. Meldrum[3] has suggested that statements on p.275 of Higgins' book hint that the author believed that atoms of some metals may be heavier than the atoms of non-metals. This evidence is not very convincing and the whole question is uncertain. Throughout his lengthy writings Higgins said so little about the relative weights of atoms of different elements that it is evident he did not attach any importance to this subject.

Higgins often failed to express himself clearly in his writings and this was one of the reasons he failed to impress his contemporaries. Another reason was that his book was probably not very widely read since it dealt with a controversy which no longer held great interest. Certainly and more importantly, the title gave no hint that the book dealt in any novel way with new views on the atomic theory.

For about 25 years after he first published his ideas, Higgins did nothing further to develop them. Despite his great originality he failed to take the crucial step that was to lead to a fruitful chemical theory of atoms.

In the meantime another scientist unaware of Higgins' work, namely John Dalton (1766–1844), had begun to think along lines that were eventually to prove more fruitful.

Dalton, now generally accepted as the originator of the chemical theory of atoms, was born in 1766 at Eaglesfield, a small village on the edge of the Lake District of England. (For recent works on Dalton see refs. 4 and 5.)

It would be difficult to imagine a greater contrast than that between the early circumstances of two founders of modern chemistry, Lavoisier and Dalton. Lavoisier was born to great wealth with all the educational advantages such wealth could give; Dalton, the son of a hand-loom weaver, for the greater part of his life earned his living as a teacher and tutor. Dalton's family belonged to the Society of Friends and it was mainly owing to this circumstance that he received any formal education—which he did up to the age of 12 only. After this, he was essentially self-taught. That he became a schoolmaster did much to aid his intellectual development, there being few better ways of learning a subject than having to teach it. With his older brother, Jonathan, he conducted a school in the neighbouring town of Kendal where he remained until he was about 23 years of age.

It would be wrong to conclude that at this period of his life, Dalton had received no educational help of any kind from others. During the latter part of the period, he became deeply interested in meteorology —his first love in science. In the preface to a book[6] which he devoted to this subject, he wrote "To one person, more particularly, I am peculiarly indebted; indeed, if there be anything new and of importance to science contained in this work, it is owing, in great part to my having had the advantage of his instruction and example in philosophical investigation". He was referring to his friend John Gough, a remarkable man who had gained repute as a mathematician and was something of a botanist despite the fact that he had been blind since the age of three. Wordsworth refers to him in his poem *The Excursion*.

Inspired by Gough's example, Dalton, on May 24, 1787 began to keep a meteorological journal and continued it until his death 57 years later. Dalton's interest in meteorology is of greatest importance in relation to his development of the atomic theory. Much of his first book *Meteorological Observations and Essays* is devoted to speculations about the nature of the atmosphere which contain the germ of his later discoveries.

Largely as a result of the influence of Gough, Dalton was appointed a tutor in Mathematics and Natural Philosophy at the New College,

References pp. 63-64

Manchester, in 1793, where he remained until 1800. It was during this period that Dalton began to teach chemistry using as textbooks Lavoisier's *Elements of Chemistry* and Chaptal's *Chemistry*. The only extensive course of lectures on chemistry that Dalton himself ever attended was a course of thirty given in Manchester in 1796 by Dr. Thomas Garnett who was later Sir Humphry Davy's predecessor at the Royal Institution.

At the invitation of the socialist, Robert Owen, Dalton joined the Manchester Literary and Philosophical Society, an institution which played a major part in Dalton's life. In it he held many offices* and before its members he read more than a hundred scientific papers. Dalton carried out his laboratory work in one of the rooms of the Society's headquarters which were destroyed by bombing in the Second World War. Much of his equipment and many of his notebooks were lost as a result of this destruction. Half-burned pages of notebooks exhibited in the University of Manchester on the occasion of the bicentenary of Dalton's birth were a grim reminder of this loss.

There has been a great deal of controversy about how Dalton originated his chemical theory of atoms. One of the first versions of the origin of the theory was given by Thomson[7] who knew Dalton personally and who, in 1807, gave the world the first textbook account of Dalton's atomic theory. Thomson visited Dalton in 1804 and talked to him about his work. Recollecting this conversation some 25 years afterwards, Thomson reported "Mr. Dalton informed me that the atomic theory first occurred to him during his investigation of olefiant gas and carburetted hydrogen" (ethylene and methane, respectively). From his analyses of the two gases, Dalton, in 1804, discovered that if one compared the amounts of hydrogen combined with a fixed weight of carbon in each case, they stood in the ratio 1:2. The implication is that Dalton was led inductively from his analyses to the law of multiple proportions, of which the above analyses afford an example, and thence, as a result of pondering over the significance of the law, to the atomic theory. While Thomson's recollection after such a long period may have been correct, this explanation is inconsistent with other evidence. It is clear from a table of atomic weights contained

---

* Secretary 1800, Vice President 1808 and President from 1817 until his death in 1844.

in a paper read on 21st October, 1803, and evidence provided by entries made in his notebook about six weeks earlier on his 37th birthday, (September 6), that Dalton had then already begun to think about the relative weights of ultimate particles.

The most recent investigations point fairly conclusively to the fact that the chemical atomic theory was not the result of Dalton's long pondering over the significance of the laws of chemical combination but, as already hinted, the result of his speculations about the nature of gases, especially those of the atmosphere.

One fact that greatly puzzled Dalton and other chemists, was how a mixture of gases of different density such as makes up the earth's atmosphere, could "constitute apparently a homogeneous mass". In order to understand Dalton's thinking about gases, it is essential to know what his mental picture of the constitution of gases was. Dalton believed, as did other scientists of the time, that gases were composed of particles each of which was surrounded by an envelope of a hypothetical fluid known as caloric. It was essentially a static mechanical model of the gas state. The particles with their surrounding envelopes were regarded as being packed close together like shot in a pile, a simile he himself used. When a gas, under constant pressure, expanded as a result of being heated, it was because the addition of heat increased the size of the envelope of caloric about each atom.

It is important to remember also that at this period, prominent chemists like Davy and Berthollet, Henry and Thomson believed that the gases of the atmosphere were chemically combined though the chemical affinity between them was thought to be very weak. The distinction between compounds and homogeneous mixtures was one that troubled many chemists at that time.

After studying the nature and varying amounts of water vapour in the atmosphere, Dalton reached the conclusion that water vapour was an "aeriform fluid" (*i.e.* gas) in its own right and that it was mechanically mixed with the remainder of the atmosphere. From this, he took the next step: namely, to conclude that all the gases of the atmosphere were mechanically mixed and not chemically combined.

To account for the homogeneity of the atmosphere he proposed, in 1801, a new theory[8] of the constitution of mixed aeriform fluids (gases) and particularly of the atmosphere. He stated[9]: "When two elastic fluids (gases) denoted by A and B are mixed together, there is no

References pp. 63-64

mutual repulsion between their particles; that is the particles of A do not repel those of B, as they do one another. Consequently, the pressure or the whole weight upon any particle arises solely from its own kind".

We see here in the notion that gas pressures arise from the repulsion of like particles, the influence of Newton. It will be recalled that he gave a mathematical explanation of Boyle's Law based on the assumption that the particles of the gas repelled one another by a force varying inversely as the distance between them. He added "Whether elastic fluids (*i.e.* gases) do really consist of particles so repelling one another is a physical question".

The assumption that particles only repelled those of their own kind was purely arbitrary and one of which Dalton eventually became so suspicious that he abandoned it. At the time he made the assumption he believed the particles of different gases were all alike in size but we shall see that shortly afterwards, in 1803, he also changed his mind about this.

Dalton used the theory of mixed gases to deduce that, in a mixture of gases, each exerts its own pressure independent of other gases.

This was concisely described by Henry in the form "every gas is a vacuum to every other gas". Nowadays, it is known as the Law of Partial Pressures and is, of course, explained quite differently in terms of the Kinetic Theory. Dalton also used his theory of mixed gases to explain the diffusion of gases one into the other.

In the course of some experiments directed to making a more accurate estimate of the amount of carbon dioxide in the atmosphere, Dalton noted that lime-water absorbed carbon dioxide from the air more readily than did ordinary water, and he was thus led to draw a distinction between chemical and mechanical absorption. He formed the opinion that "carbonic acid gas (carbon dioxide) is held in ordinary water, not by chemical affinity, but merely by the pressure of the gas . . . on the surface forcing it into the pores of the water".

This he regarded as evidence in support of his mechanical theory of mixed gases. The theory was attacked by Davy, Berthollet, Thomson and also by his friend Henry, who decided to put it to further tests. To do this, he made "an extensive series of experiments to ascertain the order of affinities of gases for water". Instead of finding any evidence of chemical affinity, his results confirmed Dalton's finding in that he showed that, at a constant temperature, the mass of a gas dissolved in water is proportional to its pressure; furthermore, that the solubilities of gases in a mixture are proportional to their partial pressures. There seemed to be no doubt that "absorption of gases by water is a purely mechanical effect."

By now Dalton's curiosity about the solubility of gases in liquids was thoroughly aroused. On October 21, 1803, he read a paper before the Literary and Philosophical Society of Manchester (*On the Absorption of Gases by Water and other Liquids*)[10].

"Why", he asked "does not water admit its bulk of every kind of gas alike? This question I have duly considered and though I am not able yet to satisfy myself completely, I am nearly persuaded that the circumstance depends upon the weight and number of the ultimate particles of the several gases. Those whose particles are lightest and single being least absorbable and the others more according as they increase in weight and complexity".

At the end of the paper on the absorption of gases which, incidentally, was not published until 1805, Dalton added significantly[11]: "An inquiry into the relative weights of the ultimate particles of bodies is a subject as far as I know, entirely new; I have lately been prosecuting this enquiry with remarkable success. The principle cannot be entered upon in this paper; but I shall just subjoin the results as far as they appear to be ascertained by my experiments". See Table 2 which is based on entries in Dalton's notebook (ref. 12, Vol. 1, pp. 244 and 248).

There is now no doubt that Dalton's decision to investigate "the relative weights of the ultimate particles of bodies" arose, as both Nash[13] and Thackray[14] have clearly shown, from a continuing endeavour to find wider experimental support for his theory of mixed gases and his speculation on the question "Why does water not admit its bulk of every kind of gas alike?"

Thus, we have the paradoxical situation that Dalton's chemical theory of atoms arose not from a consideration of chemical reactions but of a physical phenomenon—the solubility of gases in water. Half the battle in making some scientific discoveries is won by posing the right question. In the present instance, it was "What are the relative weights of the ultimate particles of bodies?"

Concerning his speculation about the relationship between particle size and solubility, Dalton later wrote: "Subsequent experience ren-

TABLE 2

OBSERVATIONS ON THE ULTIMATE PARTICLES* OF BODIES AND THEIR COMBINATIONS

Dalton's notebook, September 6, 1803.

- ◯ Hydrogen
- ⦿ Oxygen
- ⏀ Azote (nitrogen)
- ⬤ Carbone, pure charcoal
- ⊕ Sulphur

Hydrogen	1
Oxygen	5.66
Azote (nitrogen)	4
Carbone	4.5
Water	6.66
Ammonia	5
Nitrous gas (nitric oxide)	9.66
Nitrous oxide	13.66
Nitric acid (nitrogen dioxide)	15.32
Sulphur	17
Sulphurous acid	22.66
Sulphuric acid	28.32
Carbonic acid (carbon dioxide)	15.8
Oxide of carbone (carbon monoxide)	10.2

In most instances Dalton used the results of other chemists for his calculations of particle weights. *e.g.*: oxygen from Lavoisier's result for water, 85 % oxygen, 15 % hydrogen; nitrogen from Austin's analysis of ammonia (1788), 80 % nitrogen, 20 % hydrogen; carbon from Lavoisier's synthesis of carbon dioxide, 72 % oxygen, 28 % carbon.

On p. 249 of his notebook (presumably still under the date September 6) Dalton shows his formulae for a number of compounds some of which are shown below:

⏀⦿⏀	Nitrous oxide	⦿⬤⦿	Carbonic acid (carbon dioxide)
⏀⦿	Nitrous gas (nitric oxide)	⊕⦿	Sulphurous acid SO_2
⦿⏀⦿	Nitric acid (nitrogen dioxide)	⦿⊕⦿	Sulphuric acid
◯⦿	Water	◯	
⏀◯	Ammoniac (ammonia)	⬤⦿⬤	Alcohol
◯⬤	Gaseous oxide of carbon (carbon monoxide)	◯	

Table 2, the earliest known table of atomic weights differs in several respects from the first published table which appeared in 1805 (ref. 10). Besides several small changes in the values for particle weights there are several significant additions, the most important being the values for three carbon compounds.

Alcohol	15.1
Carburetted hydrogen from stagnant water (methane)	6.3
Olefiant gas (ethylene)	5.3

The last two values are the results of analyses made by Dalton during 1804.

* Dalton used the terms particle, ultimate particle and atom interchangeably though the second was the one he used most frequently.

ders this conjecture less probable". Nevertheless it had sparked off the vital question.

For the assumptions underlying the answer Dalton gave to this question, we turn again to the entries in his notebook for September 6, 1803. It is Partington's view[15] that these entries contain implicitly or explicitly the following principles:

(1) Matter consists of small ultimate particles or atoms.
(2) Atoms are indivisible and cannot be created or destroyed (Law of the Indestructibility of Matter in atomic terms).
(3) All atoms of a given element are identical and have the same invariable weight.
(4) Atoms of different elements have different weights.
(5) The particle of a compound is formed from a fixed number of atoms of its component elements (Law of Constant Proportions).
(6) The weight of a compound particle is the sum of the weights of its constituent atoms.
(7) The weight of an atom of an element is the same in all its compounds so that the composition of a compound of two elements A and C may be deduced from the composition of compounds of each with a third element B (Application of Law of Reciprocal Proportions).
(8) If only one compound of two kinds of atoms A and B is known, it is, unless there is some reason to the contrary A+B. If there is more than one compound one is A+B and the other 2 A + B or A+2B and so on. It is worth noting that at this early date, September 6, Dalton wrote N+O for nitric oxide, 2N+O for nitrous oxide and N+2O for nitrogen dioxide; also C+O for carbon monoxide and C+2O for carbon dioxide. Modern symbols have been substituted for Dalton's.
(9) Equal volumes of different gases under the same conditions of temperature and pressure *cannot* contain the same number of ultimate particles, since water vapor, the particle of which contains at least one atom of oxygen, is lighter than oxygen gas.

The essence of Dalton's contribution to the theory of atoms is contained in (3), (4), (5) and (8). By means of these assumptions, he interpreted the classification of pure substances into elements and compounds in terms of atoms and molecules as Higgins had done before him. Having posed the question what are the relative weights of

References pp. 63-64

particles of hydrogen and oxygen, for example, Dalton turned to the results of Lavoisier's analysis of water for an answer. From now on he must have proceeded, partly at least, by the hypothetico-deductive method.

One can imagine that his train of thought may have run something like this: Let us assume (a) that all atoms of hydrogen are identical in weight; (b) that all atoms of oxygen are identical in weight but different in weight from those of hydrogen. Then, if the formation of water results from an atom to atom combination of hydrogen and of oxygen to form particles of water ("compound atoms" of water Dalton called them), the ratio of the weight of oxygen to the weight of hydrogen present in any sample of water, will reveal the ratio of the weights of the ultimate particles (atoms) of oxygen and hydrogen. Moreover, the ratio of the weight of oxygen to the weight of hydrogen will be constant in all samples of water.

At this point it is worth pausing to examine assumptions (a) and (b) and to consider the way in which Dalton came to make them. On the first we have his own words:

"Whether the ultimate particles of a body [a substance] such as water are all alike, that is, of the same figure [shape], weight, *etc.* is a question of some importance. From what is known, we have no reason to apprehend a diversity in these respects: if it does exist in water, it must equally exist in the elements constituting water, namely, hydrogen and oxygen. Now it is scarcely possible to conceive how the aggregates of dissimilar particles should be so uniformly the same. If some of the particles were heavier than others, if a parcel of the liquid on any occasion were constituted principally of these heavier particles, it must be supposed to affect the specific gravities of the mass, a circumstance not known. Similar observations may be made on other substances. Therefore we may conclude that the ultimate particles of all homogeneous bodies are perfectly alike in weight, figure, *etc.* In other words, every particle of water is like every other particle of water; every particle of hydrogen is like every other particle of hydrogen".

As to assumption (b) (atoms of different elements have different weights), it should be noted that (9) implies that Dalton believed the particles of different gases were different in size and therefore different in weight. His motive for determining the relative weights of the particles of different gases was, presumably, to discover their relative sizes. Once he had pictured the particle of water as being formed by the combination of one atom of hydrogen with one atom of oxygen ((5) and (8)), the analytical figures for the chemical composition of water would immediately have suggested that the atoms of oxygen differed in weight from those of hydrogen.

John Dalton. Bronze statue in front of the John Dalton College, Manchester

On the basis of principles (3), (4) and (5) Dalton was able to conclude that compounds must be constant in composition. It is sometimes suggested that Dalton arrived at his theory the other way round, that is, in an attempt to account for or to "explain" the law of constant proportions. It should be remembered, however, that in 1803 this law was the subject of a considerable controversy. Meldrum[16] has pointed out that "the doctrine that chemical compounds have a constant composition is not a discovery made in the eighteenth century by a certain man". Many of the outstanding chemists of the latter part of that century, Lavoisier, Richter, Wenzel, Klaproth and Cavendish seem to have accepted the constancy of composition without question.

References pp. 63-64

Then suddenly at the beginning of the 19th century the validity of this law was challenged by no less a person than Berthollet (1748–1822), the leading French chemist of that day. His principal, and for some time only opponent, was Proust (1754–1826).

In support of his views Berthollet cited the variable composition of solutions, glasses and some alloys—examples of what would now be regarded as homogeneous mixtures. Proust, on the other hand made an extensive study of the composition of metal oxides, sulphides and various relatively simple inorganic substances.

The complexities of the controversy are much too great to permit any detailed analysis of it here. Some confusion arose from inaccurate analyses but probably the greatest confusion arose from the fact that materials for which variable analysis were reported consisted of variable mixtures—mixtures of two oxides (cuprous and cupric oxides), of an oxide and a hydroxide (calcium oxide and calcium hydroxide) and mixtures of normal and acid salts. Proust[17], when pressed by Berthollet to give a precise definition of a true chemical compound found it very difficult to do so and was practically driven to using constancy of composition as the essential criterion*. (See also ref. 18.)

Thus he argued: "According to our principles a compound is . . . . a privileged product to which Nature assigns fixed proportions; it is in short, a being which she never creates even between the hands of man otherwise than balance in hand, *pondere et mensura*". Dalton's great achievement lay in providing an atomic basis for the law and there is no doubt this greatly accelerated its general acceptance.

In his attempt to discover the relative weights of atoms, Dalton was obliged to make some arbitrary simplifying assumptions—a perfectly reasonable thing to do in the circumstances. These he expressed in his rule of greatest simplicity which is embodied in principle (8).

Thackray has recently pointed out that these rules are not as arbitrary as they might seem or even as arbitrary as Dalton[19] himself thought them to be when he presented them in 1808 in *A New System of Chemical Philosophy*. Thus, he notes "that Dalton did set out, in a now almost totally forgotten paper[20] the rationale that underlay his combining rules and as the following quotation clearly shows this

---

* Events were to prove that Berthollet's view had a grain of truth in it; there are substances of variable composition now called Berthollide compounds.

rationale was clearly derived from the theory of mixed gases of 1801. 'When the element A has an affinity for another B, I see no mechanical reason why it should not take as many atoms of B as are presented to it, and can possibly come in contact with it (which may probably be 12 in general) *except so far as the repulsion of the atoms of B among themselves more than a match for the attraction* of atom A. Now this repulsion begins with two atoms of B to one of A, in which case the two atoms of B are diametrically opposed (B.A.B.); it increases with three atoms of B to 1 of A in which the atoms of B are only 120° as under

$$\begin{matrix} & B & \\ & A & \\ B & & B \end{matrix}$$

It is evident from these positions that as far as the powers of attraction and repulsion are concerned, (and we know of no other in chemistry) *binary* compounds AB must first be formed in the ordinary course of things, then ternary AB_2 and so on . . . . till the repulsion of the atoms of B or A whichever happens to be on the surface of the other, refuse to admit any more'".

In 1804, Dalton analysed the two gases carburetted hydrogen (methane) and olefiant gas (ethylene) and obtained these results:

	Methane (%)	*Ethylene* (%)
Carbon	74.85	85.62
Hydrogen	25.5	14.38

Expressed as percentages, the figures are not very illuminating. It was Dalton's genius to put them on a different basis by stating the amount of hydrogen which combines with the same amount of carbon. Thus, if we use the figure 4.5 for carbon (Dalton's value for the atomic weight of this element), we obtain:

	Methane	*Ethylene*
Carbon	4.5	4.5
Hydrogen	1.51	0.75

References pp. 63-64

Dalton gave an explanation of the simple relationship that exists between the different amounts of hydrogen which combine with the same amount of carbon (2:1) in terms of the atomic theory. The results were precisely what he expected if one atom of carbon combines with two atoms of hydrogen to form methane; one atom of carbon combines with one atom of hydrogen to form ethylene.

In the course of the application of the rule of greatest simplicity to specific cases like the above (and to the oxides of nitrogen), Dalton discovered the Law of Multiple Proportions—one of his most important contributions to the atomic theory. Although the following comment by Dalton was probably made in a quite general sense, it can be applied appropriately to his own work. "Facts and experiments, however, relating to any subject, are never duly appreciated until in the hand of some skilful observer they are made to foresee the consequences of certain other operations which were never before undertaken".

The manner in which he consolidated the results of his thinking a few years after he put forward his theory is perhaps best illustrated by quoting a passage from his book[19]. This was, in fact, the first account of the theory that he published.

"Chemical analysis and synthesis go no farther than to the separation of particles one from another, and to their reunion. No new creation or destruction of matter is within the reach of chemical agency. We might as well attempt to introduce a new planet into the solar system, or to annihilate one already in existence, as to create or destroy* a particle of hydrogen[20]: All the changes we can produce, consist in separating particles that are in a state of cohesion or combination, and joining those that were previously at a distance.

Now it is one great object of this work, to show the importance and advantage of ascertaining the relative weights of the ultimate particles, both of simple and compound bodies, the number of simple elementary particles which constitute one compound particle, and the number of less compound particles which enter into the formation of one more compound particle".

* Thackray[21] has queried whether Dalton really believed that atoms were indivisible. Commenting on this passage he says "But even here Dalton did *not* state that a particle of hydrogen cannot be chemically divided." It is reasonable to believe that when Dalton used the term ultimate particle he implied indivisibility. Dalton was influenced to a considerable extent by Newton who accepted the indivisibility of the atom.

Dalton was the first to attempt to discover the relative weights of atoms by means of a study of the composition of chemical compounds. Apparently, quite independently of Dalton, Grotthuss[22] in 1806 represented the formula of water by ⊙ ○ (OH) but although he knew the combining weights of hydrogen and oxygen he, like Higgins, failed to take the crucial step. Grotthus[23] wrote: "It is remarkable how near I was to Dalton's discovery relating to the weights of atoms. The diagrams in Dalton's *System* are exactly the same as I had used much earlier. I admit, however, that the beautiful idea of determining the relative weights of atoms from the composition of bodies had not occurred to me". (See also ref. 2, p. 131.)

Although Dalton's main objective was to determine the true relative weights of atoms, ironically enough, he never succeeded in doing this. He was unable to find any method of checking the fundamental assumptions of the rule of greatest simplicity. On the basis of this rule, he assumed, as already pointed out, that the molecule of water was formed by the combination of one atom of hydrogen with one atom of oxygen and wrote its formula as shown in Table 1.

Since there are, in fact, two atoms of hydrogen in a molecule of water, Dalton's calculation of the atomic weight of oxygen gave a result which was half its true value. Similarly, with ammonia which he represented by the formula ⊙ ○ (NH in modern symbols); we now know there are three atoms of hydrogen in a molecule of ammonia (NH_3) and this means that his value for the atomic weight of nitrogen is about one-third of its true value. He represented methane by the formula, CH_2 (correct formula is CH_4) and consequently his value for the atomic weight of carbon is about half its true value. None of the analytical results used by Dalton in calculating these atomic weights was very accurate. Had they been accurate, his atomic weights would have been exactly one-half or one-third instead of about one-half or one-third the true values. It will be recalled that in the first instance, Dalton calculated relative weights of particles in an effort to vindicate his theory of mixed gases and the solubility of gases in water. It was only gradually that he realised the importance of his work for chemistry as a whole. In this he was undoubtedly assisted by Thomson who was much more of a chemist than Dalton.

Thomson was always very careful to give Dalton full credit for his theory. Thus he later said "I call it the Daltonian theory because I

References pp. 63-64

consider it belongs to Mr. Dalton; because he first suggested it to me and set me thinking on the subject; and, of course, everything here stated originated from him, either directly or at least indirectly".

Thomson made explicit a fundamentally important step, the inspiration for which must almost certainly have come from Dalton, when he attempted to determine the relative numbers of atoms in compounds containing more than two different kinds of atoms. At that time it was known that cane sugar and oxalic acid both contained carbon, hydrogen and oxygen, though each in different proportions and Thomson became interested to find out why these compounds differed from one another.

From a knowledge of the composition of binary compounds of assumed formulae, AB, AB_2 etc., Dalton had calculated atomic weights (oxygen 6, nitrogen 5, carbon 4.5, hydrogen 1). Thomson[24], using these values for the atomic weights, put the calculation into reverse, that is, from a knowledge of the percentage composition of oxalic acid and the above atomic weights, he proceeded to calculate the relative numbers of the different atoms present in oxalic acid. This he did by dividing the percentage of each element by its atomic weight and then converting the resulting figures into integral ratios. The calculation based on modern atomic weights for anhydrous oxalic acid is shown in the footnote*.

From the data then available, Thomson concluded that a molecule of oxalic acid contained 4 atoms of oxygen, 3 of carbon and 2 of hydrogen, and wrote the formula for oxalic acid thus 4 w + 3 c + 2 h (w = oxygen, c = carbon and h = hydrogen). This is one of the earliest empirical formulae ever written for a molecule containing atoms of more than two elements and in which the symbols have a quantitative

---

* Percentage composition of anhydrous oxalic acid

$H = 2.23 \%$ $C = 26.68 \%$ $O = 71.08 \%$

$$\frac{2.23}{1.008} = 2.22 \qquad \frac{26.68}{12.011} = 2.22 \qquad \frac{71.08}{16.00} = 4.44$$

Converted to integral ratios these become 1 : 1 : 2. Thus the empirical formula is HCO_2. The empirical formula gives the relative numbers of atoms of different kinds of in a molecule; the molecular formula gives absolute number. The molecular formula for oxalic acid is $H_2C_2O_4$. This point is further elaborated on p. 145, Chapter IX.

sense*. That the formula is incorrect is of little account now; its importance lies in the fact that it embodied a new use of analytical data.

This was a revolutionary step—an entirely new way of describing a substance and an important advance in the long process of unravelling the structure of matter.[25]

Another revolutionary change in outlook consequent on the work of Dalton, stemmed from the fact that his theory embodied the second major modification of the Greek atomic theory. This theory held that all atoms were formed from the same universal matter. In Dalton's theory, the atoms of each element are distinctive species of matter or to put it another way the different kinds of atoms are irreducibly different. This implication of the Daltonian theory amounted to an abandonment of the belief in the unity of matter. The essential step in this abandonment was taken by Lavoisier when he defined a chemical element. Dalton translated it into atomic terms. Those who opposed the chemical theory of atoms in the early nineteenth century sometimes did so because of a refusal to abandon belief in the unity of matter**.

It is appropriate here to consider briefly the question raised earlier—whether Dalton was, in fact, the originator of the chemical theory of atoms.

Not long after Dalton put forward his views, he was challenged by Higgins who claimed that he was the originator of the theory

In a book[27] published in 1814, Higgins wrote—"It will be found that Dalton has not done justice to my doctrine with all his ingenuity; and his attempt to weigh atoms no matter how, or whether he is correct or not, gives him no claim whatever to the system which I established several years before he or Dr. Thomson were known as chemical writers".

---

* Translated into modern symbols Dalton's formula for alcohol derived in 1803 was C_2H_2O (see p. 52). To have arrived at this he must have used a process similar employed by Thomson.

** Foremost among those who opposed the theory on these grounds was Humphry Davy (1778–1829) who despite the fact that he had discovered six metals was, for a time at least, so doubtful of their elementary nature that he was moved to declare that "To enquire whether the metals be capable of being decomposed is a grand object of true philosophy."[26]

References pp. 63-64

From this passage it is evident that even at this late date Higgins failed to grasp the importance of a knowledge of the relative weights of the atoms of different elements.

In his early days, said Higgins, "Philosophers were then eager to attribute the merit of a discovery to its rightful owner, not to appropriate it to themselves or others. But now in the vale of my life, I am obliged to rescue the labors of my youth from the claims of those who have adopted them without ceremony.... The subject is not now confined to the decision of a few individuals but is laid before a grand tribunal and it rests with them to give a verdict".

The grand tribunal was slow in reaching its final verdict. As recently as 1951, Higgins' claim was stoutly defended in the columns of *Nature* by no less a person than the late Professor Soddy[28].

Like most controversies, there was something to be said on both sides.

The part played by Higgins has already been described. However, despite all his contributions to the atomic theory, Higgins missed a crucial point, the relationship between atomic weights and chemical composition. In a recent, thoroughly documented and detailed study, Partington and Wheeler have given an impartial account of the controversy. They sum up their verdict thus: "The theory that the atom of every element has a characteristic weight belongs unquestionably to John Dalton, and it is Dalton's theory and not Higgins' which is the basis of the modern atomic theory".

The essence of Dalton's originality was his attempt to discover the relative weights of atoms from study of the composition of chemical compounds. From the subsequently successful solution of this problem came many of the great advances in chemistry made in the 19th century.

For some years after the publication of *A New System of Chemical Philosophy* in 1808, Dalton's theory, as might be expected, received a mixed reception. In London Davy was very sceptical about it to say the least. In Paris Berthollet would have none of it, but in Glasgow and Edinburgh where a few years previously Dalton had also lectured, his theory was given such an enthusiastic reception that he dedicated his book to the professors of the universities in those cities. His greatest supporters were Avogadro in Turin and Berzelius in Stockholm. The last two chemists did more than any other to advance the atomic theory in the first half of the 19th century.

In his lifetime, Dalton received many honors. "For the development of the chemical theory of Definite Proportions, usually called the Atomic Theory and for his various other labors and discoveries in physical and chemical science" he was awarded a Royal Medal of the Royal Society. The French Académie des Sciences accorded him the high honor of Foreign Associate in 1830. On the occasion of the second meeting of the British Association for the Advancement of Science held at Oxford in 1831, the University conferred on Dalton, the honorary degree of D.C.L. But for his failing health, Dalton would have been elected President of the British Association when it met in Manchester in 1842. He died two years later.

REFERENCES

1 W. HIGGINS, *A Comparative View of Phlogistic and Anti-Phlogistic Theories*, London, 1789.
2 T. S. WHEELER AND J. R. PARTINGTON, *The Life and Work of William Higgins, Chemist*, Pergamon Press, Oxford, 1960, including reprint of Ref. 1.
3 A. N. MELDRUM, Development of the Atomic Theory, *Memoirs, Manchester Lit. Phil. Soc.*, 55 (1910) No. 4, p. 14. This is one of a series of seven papers on the subject.
4 D. S. L. CARDWELL, *John Dalton and the Progress of Science*, Manchester University Press, 1968. Papers presented to a Conference held to mark the bicentenary of Dalton's birth.
5 F. GREENAWAY, *John Dalton and the Atom*, Heinemann, London, 1966.
6 J. DALTON, *Meterological Observations and Essays*, London, 1793.
7 T. THOMSON, *A System of Chemistry*, Edinburgh, 1807.
8 J. DALTON, *Nicholson's J.*, 5 (1801) 241.
9 J. DALTON, *Manchester Mem.*, 5 (1802) 538.
10 J. DALTON, W. H. WOLLASTON AND T. THOMSON, *Foundations of the Atomic Theory*. Comprising papers and extracts. Alembic Club Reprint No. 2, Livingstone Ltd., Edinburgh, 1948.
11 J. DALTON, *Manchester Mem.*, Second Series, 1 (1805) 271.
12 H. HARTLEY, *John Dalton (F.R.S.), (1766–1844) and the Atomic Theory* – A Lecture to commemorate his bicentenary, Proc. Roy. Soc. A, 300 (1967) 291.
13 L. NASH, The Origin of Dalton's Chemical Atomic Theory, *Isis*, 47 (1956) 101.
14 A. W. THACKRAY, The Emergence of Dalton's Chemical Atomic Theory, *Brit. J. Hist. Sci.*, 3 (1966) 1. See also by the same author: The Origin of Dalton's Atomic Theory, Daltonian Doubts Resolved, *Isis*, 57 (1966) 35.
15 J. R. PARTINGTON, *A History of Chemisty*, Vol. 3, Macmillan and Co., London, 1964, p. 784.
16 A. N. MELDRUM, *Manchester Mem.*, 54 (1910) No. 7, p. 1.
17 L. J. PROUST, *J. de Phys.*, 63 (1806) 364.
18 A. W. THACKRAY, *Brit. J. Hist. Sci.*, 3 (1966) 650.
19 J. DALTON, *A New System of Chemical Philosophy*, Part 1, R. Bickerstaff, London, 1808, p. 216. Facsimile edition, W. Dawson and Sons Ltd., London, undated.
20 J. DALTON, *Nicholson's J.*, 29 (1811) 147.

21 A. W. THACKRAY, *Isis*, 57 (1966) 37.
22 T. GROTTHUS, *Ann. Chim.*, 58 (1806) 54.
23 T. GROTTHUS, *Schweigger's J. Chem.*, 20 (1818) 225.
24 T. THOMSON, *Phil. Trans.*, 98 (1808) 63. See also Alembic Club Reprint No. 2, Livingstone Ltd., Edinburgh, 1948, p. 41.
25 T. S. KUHN, *The Structure of Scientific Revolutions*, Chicago University Press, 1962, p. 130.
26 H. DAVY, *The Collected Works of Sir Humphry Davy*, 9 Vols, London 1839–40, Vol. IX, p. 385
27 W. HIGGINS, *Experiments and Observations on the Atomic Theory*, London, 1814.
28 F. SODDY, *Nature*, 167 (1951) 735.

CHAPTER 5

Avogrado partly overcomes a limitation of Dalton's theory

In retrospect it is not surprising that the early history of the chemical theory of atoms should have been closely linked with a growing understanding of matter in the simplest of its three states—the gaseous state. It will be recalled that the theory originated in Dalton's speculations about the physical behaviour of gases, more particularly about their diffusion and their solubility in liquids.

The next stage in the development of the theory which was ultimately to provide an answer to the problem of finding the true relative weights of atoms of different elements, was profoundly influenced by investigations of the chemical combination of gases with one another. But before taking up this subject, it will be worthwhile to examine briefly the state of knowledge of gases prevailing at the beginning of the 19th century. By then, the uniformity of the behaviour of gases under changing conditions of temperature and pressure, the outward sign of the relative simplicity of the gaseous state, had been amply demonstrated. When Boyle and others in the 17th century investigated the relationship between the volume and pressure of a gas at constant temperature, they used air as their experimental material. In the succeeding century, after the different common gases had been discovered, Boyle's law was generalised thus: the volume occupied by any species of gas, is, at constant temperature, inversely proportional to its pressure. Similarly, the effect of change of temperature on the volume of a gas at constant pressure, first investigated in 1702 by Amontons using air, was later generalised by Charles, Dalton and Gay-Lussac. In 1802, Gay-Lussac[1] wrote:

"The experiments which I have now reported and which have all been made with great care prove incontestably that atmospheric air, the gases oxygen, hydrogen, nitrogen, nitrous oxide, ammonia, muriatic acid [hydrogen chloride] and carbonic acid [carbon dioxide] all have the

References p. 83

same expansion between the same degrees of heat [*i.e.* the same temperatures]".

Gay-Lussac (1778–1850) belonged to a small private scientific society founded by Berthollet and known as the Société d'Arcueil. The Society numbered among its members some of the most eminent French chemists of the day and met for discussions in Berthollet's country house which was situated in the village of Arcueil on the outskirts of Paris. Berthollet who was the leader of the group and who undoubtedly influenced Gay-Lussac's outlook did not accept the constancy of the composition of compounds or the Daltonian atomic theory. Neither did Laplace, another famous member of the Society. This then was the climate of scientific thought in which Gay-Lussac discovered one of the most important laws of chemical combination—that relating to the volume relationship in the chemical combination of gases. Gay-Lussac[2] was probably the first to possess some insight into the origin of the relative simplicity of the gaseous state, for he wrote:

"Substances, whether in the solid, liquid or gaseous state, possess properties which are independent of cohesion; but they also possess others which appear to be modified by this force (so variable in its intensity) and which no longer follow any regular law. The same pressure applied to all solid or liquid substances would produce a diminution in volume differing in each case, while it would be equal for all elastic fluids (gases). Similarly heat expands all substances; but the dilation of liquids and solids have hitherto presented no regularity and it is only those of elastic fluids which are equal and independent of the nature of each gas. The attraction of molecules in solids and liquids is therefore the cause which modifies their special properties; and it appears that it is only when the attraction is entirely destroyed, as in gases, that bodies under similar conditions obey simple and regular laws". Although Gay-Lussac is not explicit on this point, it would seem to have been his view that the attraction between molecules in a gas vanishes because their distance from one another is greater than in solids and liquids.

That the molecules of a gas are separated from one another by distances which are large in comparison with their own dimensions is evident from their compressibility and the large volume change that occurs in the transition from the solid or liquid to the gaseous state. As an example of the latter known about this time, one may cite Watt's discovery made some years earlier that one cubic inch of water produces, at the boiling point, approximately one cubic foot of steam.

That gases liberated from solids, for example, carbon dioxide from limestone, occupy a very much greater volume than the solids from which they are liberated, as Black discovered, points to the same conclusion; molecules in the gaseous state are widely separated from one another.

Although Lavoisier had demonstrated the role of oxygen of the air in combustion and respiration and had obtained an approximate knowledge of the proportions of its two main components, many questions about the atmosphere remained unanswered. In the opening years of the nineteenth century chemists were studying the composition of air by more accurate methods in an endeavour to discover whether it was constant and independent of location either on the earth's surface or in altitude.

In an attempt to answer the latter question, Gay-Lussac on 16th September, 1804, made a balloon ascent of about 7 km above Paris and collected samples of air for analysis. Shortly after the ascent, he and M. Humboldt began analysing the samples by exploding in an eudiometer a known volume of air with a known volume of hydrogen and measuring the resulting contraction in volume. The accuracy of the method depended on the accuracy with which the ratio of the combining volume of hydrogen and oxygen was known. This they set out to determine as accurately as they could by varying the proportions of the two gases in different experiments. In one series of experiments they placed 100 parts by volume of oxygen and 300 parts by volume of hydrogen in an eudiometer and exploded the mixture. As a result of twelve experiments, they found that the mean value of the residue of hydrogen was 101.3 parts. The contraction represents the volume of hydrogen and oxygen removed in the form of water. After making a small correction for the impurity of their oxygen they found that the ratio, hydrogen to oxygen, was 199–89 : 100. "Thus it seems to us proved that 100 parts by volume of oxygen requires almost exactly (à très peu près) 200 parts of hydrogen to saturate it".

Gay-Lussac was by no means the first to discover that the ratio of the combining volumes of hydrogen and oxygen approximates to 2:1. The results quoted by Cavendish in 1784 may be used to calculate a ratio of 2.02:1. Five years later, Higgins (see Chapter 4, ref. 1) stated, without quoting any supporting experimental data: "2 cubic inches of light inflammable air (hydrogen) require but 1 of dephlogisticated air (oxygen) to condense them". The French chemists, Monge (1786) and Sequin and his colleagues (1791) also obtained similar values— 1.95:1 and 2.05:1. Dalton's first published value for this ratio was 1.87:2 but he later obtained a value of 1.97:1.

What made Gay-Lussac, some 4 years later, ask the question: do other gases combine in this manner, that is, in volumes which bear a simple ratio to one another? While no definite answer can be given to this question, a clue is provided by his attitude to research. He was dominated by the urge to discover laws of Nature. He had already made an important discovery about the universality of the rate of thermal expansion of gases. He himself said: "Laws are necessarily derived from the observation of a large number of facts; but if one were not animated with the desire to discover laws they would often escape the most enlightened attention". This, however, is only part of the truth. To some extent, the discovery of the law of combining volumes was the result of a "lucky accident"[3].

In one of their many experiments made in the hope of isolating the element fluorine, Gay-Lussac and Thenard heated a mixture of calcium fluoride and boric acid in an iron tube. The chemical reaction that ensued resulted in the liberation "fluoric acid" (boron trifluoride). On coming into contact with air, this gas produced dense white fumes that reminded them of those produced by hydrogen chloride and ammonia. Further experiments showed that the fuming of "fluoric gas" was the result of its interaction with the moisture present in the air. Fluoric gas also fumed in the presence of ammonia but the reaction here was quite different.

"Suspecting, from the exact ratio of 100 of oxygen to 200 of hydrogen which M. Humboldt and I had determined for the proportions of water that other gases might also combine in simple ratios, I made the following experiments. I prepared fluoric, muriatic and carbonic acid gases and made them combine successively with ammonia gas. 100 parts of muriatic acid saturate precisely 100 parts of ammonia gas"[2] and so with the other two gases. Gay-Lussac also confirmed his hunch that simple ratios were general for gaseous combinations from the work of other investigators.

For example, he used the results of Davy on the gravimetric analysis of the oxides of nitrogen to confirm his conclusion. Davy had found the following proportions by weight:

Oxide	Nitrogen	Oxygen
Nitrous oxide	63.30	36.70
Nitric oxide	44.05	55.95
Nitrogen dioxide	29.50	70.50

Reducing these proportions to volumes Gay-Lussac found:

Oxide	Nitrogen	Oxygen
Nitrous oxide	100	49.5
Nitric oxide	100	108.9
Nitrogen dioxide	100	204.7

"The first and last of these proportions differ only slightly from 100 to 50 and 100 to 200; it is only the second which diverges somewhat from 100:100. The difference, however, is not very great and is such as we might expect in experiments of this sort; and I have assured myself that it is actually nil".

Gay-Lussac[2] summed up his results:

"I have shown in this Memoir that the compounds of gaseous substances with each other are always formed in very simple ratios, so that representing one of the terms by unity the other is one, two or at most three. These ratios by volume are not observed with solid or liquid substances, nor when we consider weights, and they form a new proof that it is only in the gaseous state that substances are in the same circumstances and obey regular laws..... The apparent contraction of volume suffered by gases is also very simply related to the volume of one of them, and this property is likewise peculiar to gaseous substances".

The simplicity of the volume relations discovered by Gay-Lussac strongly suggested that the generalization must have some fundamental importance. The 1:1 and 1:2 ratios were reminiscent of Dalton's view of the combination of atoms to form binary and ternary compounds. Gay-Lussac's law suggested that equal volumes of different elementary gases contained the same (or simply related) numbers of atoms. Dalton examined this possibility but rejected it though not without a seemingly good reason. Some of the known facts that worried Dalton were:

(i) Although an atom of hydrogen is added to an atom of oxygen in the formation of water (so he thought), the density of water vapour is less than that of oxygen.
(ii) Combination of hydrogen with nitrogen yielded ammonia which is less dense than nitrogen.
(iii) Carbon monoxide is less dense than oxygen although the former is a compound of oxygen plus carbon.

To Dalton these facts suggested that there could not be equal numbers of particles in equal volumes of different gases.

Believing in the pile of shot theory of the structure of gases and in the idea that the sizes of the particles of different gases with their attendant atmospheres of caloric must be different, Dalton found it impossible to accept the view that equal volumes of gases under the same conditions of temperature and pressure contained the same number of particles.

Let us take another argument used by Dalton. Consider the following reaction as Dalton would have thought about it.

$$\begin{array}{ccc} N & + \quad O & = NO \\ 1 \text{ vol.} & 1 \text{ vol.} & 2 \text{ vol.} \\ 1 \text{ atom} & 1 \text{ atom} & 2 \text{ compound atoms} \end{array}$$

When testing the hypothesis that equal volumes of gases (under the same conditions of temperature and pressure) contain equal numbers of particles, Dalton[4], assuming it to be true, would have said—one atom of nitrogen combines with one atom of oxygen to form two compound atoms of nitric oxide. This implies that to make one "compound atom" of nitric oxide half atoms would be necessary and Dalton would have none of this. "The truth is" he said, "I believe that gases do not unite in equal or exact measures in any one instance; when they appear to do so, it is owing to the inaccuracy of our experiments. In no case perhaps is there a nearer approach to mathematical exactness than in that 1 measure of oxygen to 2 of hydrogen; but here the most exact experiment I ever made, gave me 1.97 hydrogen to oxygen".

The argument may be put another way. Let us suppose there are 1000 atoms in a given volume of nitrogen and one thousand atoms of oxygen in the same volume. These atoms are allowed to combine atom to atom to form 1000 molecules of nitric oxide. Experiment shows that these 1000 molecules occupy two volumes—that is, 500 molecules per unit volume which is contrary to the hypothesis that equal volumes contain the same number of particles.

There seems to have been a streak of obstinacy in Dalton's character. He never accepted the results of Gay-Lussac's experiments and consequently he made no important further progress towards advancing his objective of determining the true relative weights of atoms. Furthermore, he failed to understand the solution of the problem of determining true relative weights of atoms when it was found about three years later.

The fundamental importance of Gay-Lussac's Law of Combining Volumes was first grasped by the Italian scientist, Avogadro. Avogadro (1776–1856) began his professional life as a lawyer having in 1796 been admitted to the degree of doctor of ecclesiastical law. However, his real interests lay in mathematics and physics and after studying these subjects for a number of years, he became Professor of Mathematics and Physics at the Royal College at Vercelli. He was essentially a mathematical physicist and, like Dalton before him, he used the experimental results of other scientists—in this instance those of Gay-Lussac—when first elaborating his theories. His picture of a gas was like that of Gay-Lussac insofar as he seems to have regarded the particles as widely separated. "At such a distance that their mutual attractions cannot be exercised"—was a phrase he constantly used.

There seemed to him to be two possible explanations of Gay-Lussac's Law of Combining Volumes: "The first hypothesis to present itself in this connection, and apparently even the only admissible one, is the supposition that the number of integral molecules (molecules, in modern terminology) in any gas is always the same for equal volumes or always proportional to the volume". The second possibility was that equal volumes of different gases contain numbers of molecules that are in the ratio of small integers.

Avogadro found it impossible to believe that equal numbers of molecules of different gaseous substances, widely separated as they are, could occupy volumes exactly in the ratio of small integers. Thus,

References p. 83

he adopted the former hypothesis, it being understood that all gaseous volumes were compared under the same pressure and temperature.

Like other scientists of his day, Avogadro accepted the caloric theory of heat and thus he believed that each gaseous particle was surrounded by a layer or an atmosphere of caloric. His model of a gas seems to have been like that of Dalton in that it was essentially static. But it differed from Dalton's in one important respect, namely that the vol-

Amadeo Avagadro

ume of space occupied by a particle and its envelope of caloric was the same for all gases. Avogadro sometimes expressed his hypothesis in an alternative form: under the same conditions of temperature and pressure the particles of all gases are at equal distances from one another. This point is of sufficient importance to warrant quoting his own words[5]:

"it is very well conceivable that the molecules of gases being at such a distance that their mutual attraction cannot be exercised, their varying attraction for caloric may be limited to condensing a greater or smaller quantity around them, without the atmosphere formed by this fluid having any greater extent in the one case than in the other, and, consequently, without the distance between the molecules varying; or in other words, without the number of molecules contained in a given volume being different.... But in our present ignorance of the manner in which this attraction of the molecule for caloric is exerted we should rather be inclined to a neutral hypothesis, which would make the distance between the molecules and the quantities of caloric vary according to unknown laws, were it not that the hypothesis we have just proposed is based on the simplicity of the relation between the volumes of gases on combination, which would appear to be otherwise inexplicable".

In other words, Avogadro decided that different quantities of caloric did not affect the distance between the molecules of a gas and therefore did not affect the number of molecules in a given volume. He does not say at what distances apart molecules would cease to exercise their mutual attraction. Ampere who a few years later independently proposed the same hypothesis as Avogadro seems to have regarded these distances as many times greater than the dimensions of the molecules themselves.

One of the most important consequences of the hypothesis was that it led Avogadro to recognize the existence of polyatomic (diatomic) molecules in the common elementary gases.

The line of reasoning he used in reaching this conclusion may be illustrated by considering a general problem. Let us take the case of a gas A, unit volume of which combines with x volumes of gas B (where x may be 1, 2 or 3) to form AB_x. Assume that there are n particles in unit volume of any gas.

If each particle of AB_x contains, as indicated, one atom of A then there cannot be more than n particles of product. Consequently there should, if Avogadro's hypothesis holds good, not be more than unit volume of the product AB_x. In general this does not occur.

Avogadro noticed that, almost without exception, where one volume of gas A combined with one or more volumes of gas B, the volume of the product was double that of the volume of A. He therefore, concluded that the particle of A must be a composite of two particles or as we would now say diatomic.

References p. 83

The first reaction he examined was that between oxygen and hydrogen to form water about which he wrote:

"The volume of water in the gaseous state is, as M. Gay-Lussac has shown, twice the volume of the oxygen which enters into it or what comes to the same thing equal to that of the hydrogen instead of being equal to that of the oxygen. But a means of explaining facts of this type in conformity with our hypothesis presents itself naturally enough: we suppose namely that the constituent molecules [molecules] of any simple gas whatever (*i.e.* molecules which are at such a distance from each other that they cannot exercise their mutual action) are not formed of a solitary elementary molecule [atom] but are made up of a certain number of these molecules [atoms] united by attraction to form a single one; and further that when molecules of another substance unite with the former to form a compound molecule, the integral molecule [molecule] which should result (H_4O_2 in the case of water) splits up into two or more parts composed of half the number of elementary molecules [atoms] going to form the constituent molecules [molecules] of the first substance . . ."*

If Avogadro had used modern symbols for atoms of the elements and equations, he would probably have represented the reaction between oxygen and hydrogen thus:

$$O_2 \;+\; 2H_2 \;\rightarrow\; [H_4O_2] \;\rightarrow\; 2H_2O \qquad (1)$$

1 vol.   2 vol.   2 vol.   2 vol.

n particles $2n$ particles n particles $2n$ particles

To continue with his explanation:

"On reviewing the compound gases most generally known, I find only duplication of the volume relatively to the volume of that one of the constituents which combines with one or more volumes of the other**. We have already seen this for water. In the same way, we know that the volume of ammonia gas is twice that of the nitrogen which enters it. Finally nitrous gas [nitric oxide] which contains equal volumes of nitrogen and oxygen has a volume equal to the sum of the two

---

* The above passage is a translation taken from Alembic Club Reprints No. 2. The words in square brackets are thought to be the modern equivalents of Avogadro's terms.

** He did observe at least one exception to this rule later but it does not affect the argument here.

constituent gases, that is to say, double each of them. Thus in all these cases there must be a division of the molecule into two*; but it might be possible in other cases the division might be into four, eight, *etc.*". Having established this point, Avogadro goes on to say "the integral molecule of water will be composed of a half-molecule of oxygen with one molecule of hydrogen or what is the same thing, two half molecules of hydrogen". Put another way this becomes: a molecule of water is composed of one atom of oxygen and two atoms of hydrogen.

Thus, in order to reconcile Gay-Lussac's law with the equal number of particles in equal volumes hypothesis, Avogadro doubled the number of atoms in a molecule. Dalton saw the only solution as that of halving the atoms to which, he of course, would not agree.

Nature had, as one writer put it, "played a cruel trick on the chemists by making the common gaseous elements diatomic". It has been said[9] that Avogadro was the first man in the world to know that water is H_2O. Curiously enough, in none of the three papers in which he elaborated his hypothesis did he use any symbols for atoms of elements, nor did he write a molecular formula for water, but he did imply, as already pointed out, that the molecule of water consisted of two atoms of hydrogen and one atom of oxygen. His first paper appeared in 1811, 2 years before Berzelius introduced the modern chemical symbols for atoms of the elements. Avogadro could, of course, have used Dalton's symbols but he did not do so. He became acquainted with Dalton's work through reading the account given by Thomson in his *System of Chemistry* (Chapter 4, ref. 7). In this account Thomson uses Dalton's symbols for oxygen, hydrogen and nitrogen and writes Dalton's formulae for water and ammonia (⊙○ and ⊕○). Even in a very long paper published in 1821 summarising his work on molecular weights Avogadro wrote no formulae or equations nor did he summarise the results of his atomic weights in tabular form.

There is no doubt that if Avogadro had used appropriate symbols for atoms and molecules his ideas would have been much more readily understood.

Gregory[6], a historian of chemistry, as recently as 1931, criticised

---

* Avogadro sometimes uses the expression "there must be a division of the molecule into two" when it is not clear which of two possibilities is meant (O_2 or H_4O_2), but there is little doubt that he is here referring to the diatomic molecules O_2.

References p. 83

Avogadro's work saying "Thus his Principle was ineffectively presented because it was neither correlated uniquely with experimental data nor clearly and distinctly stated nor purified from distractive ideas".

The passages from Avogadro's first paper just quoted do not justify these strictures except possibly on one score, namely, that of a possible lack of clarity of meaning arising from Avogadro's system of naming particles of different kinds. This will not be evident in these particular passages because the difficulty has been obviated by placing in parentheses the modern terms.

The question of nomenclature is one that needs examination in some detail because it has often been suggested that obscurities in nomenclature could have responsible, at least in part, for the inordinately long delay in the acceptance of Avogadro's fundamentally important ideas.

Like most continental scientists of the day, Avogadro did not use the term atom. Not once does it appear in either of the two main papers that appeared in 1811 and 1814 respectively. Instead, he used the term molecule or elementary molecule to mean atom. He did use the term atom in his later papers.

The modern equivalents of the terms used by Avogadro* shown in the footnote on this page are taken from the Alembic Club reprint on the 1811 paper.

Had Avogadro always used the term molecule with an appropriate adjective, there could have been little serious criticism of the clarity of his exposition. However, since it is generally believed that one of the main achievements of Avogadro was that he drew a distinction between atom and molecule in the gaseous state, it was probably almost a fatal weakness in his exposition of the subject that he should have used the same word molecule to mean *either* atom or molecule.

In the next two passages which describe the second major consequence of Avogadro's hypothesis, translations of his original terms have been retained for the purpose of giving some idea of the difficulties

---

* *Molécule* (translated "molecule") without qualification means in modern chemical phraseology either atom or molecule. *Molécule intégrante* (translated "integral molecule") means molecule in general but is usually applied only to compounds. *Molécule constituante* (translated "constituent molecule") is employed to denote the molecule of an elementary substance. *Molécule élémentaire* (translated "elementary molecule") stands for an atom of an elementary substance.

that may have confronted chemists of the day when reading his papers. A careful reading of the first of the two passages will reveal a situation which could have greatly puzzled his contemporaries though his meaning is in the light of hindsight, understandable with a little effort.

"Setting out from this hypothesis (equal volumes, equal numbers of molecules), it is apparent that we have the means of determining very easily the relative masses of molecules of substances available in the gaseous state, and the relative number of these molecules in compounds; for the ratios of the masses of the molecules are the same as those of the densities of the different gases at equal temperatures and pressures; and the relative number of molecules in a compound is given at once by the ratio of the volumes of the gases that form it. For example, since the numbers 1.10359 and 0.07321 express the densities of oxygen and hydrogen relative to air as unity, and the ratio of these two numbers represents the ratio of the masses of equal volumes of these two gases, it will also represent on our hypothesis the ratio of the masses of the two molecules. Thus the mass of the molecule of oxygen will be about 15 times that of the molecule of hydrogen or more exactly 15.074:1".

In this passage the word molecule has for the most part its modern meaning but in one phrase at least it could refer to atoms. Where Avogadro speaks about "the relative number of molecules in a compound" he almost certainly means the relative numbers of atoms in a molecule. The difficulty of deciding which meaning to accept is sometimes increased by the fact that since the molecules of both oxygen and hydrogen are diatomic, every statement about the relative weights of their constituent atoms also applies to the relative weights of the molecules. To put the matter another way, when Avogadro says the ratios of the masses of the particles of oxygen to those of hydrogen is 15:1, he may be referring to either the ratios of the masses of diatomic molecules or those of single atoms.

There can be little doubt that Avogadro is referring to atoms when he compares his results with those of Dalton: "thus Dalton supposes that water is formed from hydrogen and oxygen, molecule to molecule*. From this, and from the ratio by weight of the two components, it would follow that the mass of the molecule [atom] of oxygen would

---

* That is, atom to atom as Dalton clearly shows by his diagrams.

be to that of hydrogen as 7½: 1 nearly, or according to Dalton's evaluation 6:1. This ratio on our hypothesis is, as we saw twice as great, namely, 15:1. As for the molecule of water, its mass ought to be roughly expressed by $15+2 = 17$ (taking for unit that of hydrogen)* if there were no division of the molecule into two [*i.e.* if it were H_4O_2] but on account of this division it is reduced to half 8½ or more exactly 8.537 as may be found directly by dividing the density of aqueous vapour (0.625, Gay-Lussac) by the density of hydrogen (0.0732). This mass only differs from 7, that assigned by Dalton, by the difference in the values for the composition of water; so that in this respect Dalton's result is approximately correct from the combination of two compensating errors—the error in the mass of the molecule [atom] of oxygen and his neglect of the division of the molecule".

It is of interest, in view of the later history of atomic weight determinations to compile a list of the atomic weights based on a study of the relative densities of elementary gases recorded in Avogadro's first paper (1811). Those from his 1821 paper[7] are shown in the second column.

	At. wt. (1811)	At. wt. (1821)
H	1	1
C	11.36	11.36
N	13.3	14.0
O	15.07	16.03
Cl	33.9	36.12

Though not very accurate, they are all reasonably close to the modern values.

The French physicist, Ampère, is usually credited with the independent proposal of the "equal volume equal number of molecules" hypothesis. Since for about the next 40–50 years the credit for the hypothesis was nearly always given to Ampère, it is worthwhile briefly to examine the nature of his publication of it. It took the form of a letter to Berthollet and was published[8] 3 years after Avogadro's first paper.

He explained to Berthollet that he was writing a book which, though nearly finished, he could not complete because of the pressure of work.

* In modern terms $H_2 = 1$.

In compliance with a request from Berthollet, Ampère sent him an extract from his book. One of the first things one notices is the nomenclature Ampère adopted. Thus he used the term molecule to mean an atom and the word particle to mean a molecule which may have been somewhat confusing to his contemporaries. Ampère[8] stated the hypothesis thus:

"The particles of all gases, simple or compound, are at an equal distance one from the other. The number of particles on this supposition, is proportional to the volume of the gas".

In a footnote to this statement of the hypothesis, Ampère explained that since writing his treatise he had learned (presumably as a result of reading the *Journal de Physique*) that Avogadro had used his idea "to determine the proportions of elements in chemical combinations". Ampère gave no numerical examples of atomic weights. Like Avogadro he did not write molecular formulae but he did state that the molecules of hydrogen, oxygen, nitrogen and chlorine are all tetra-atomic (using modern terms) but without giving any satisfactory justification for doing so. He stated the composition of the water molecule in terms of four atoms of hydrogen and two atoms of oxygen.

It is of course possible to satisfy the volume relationship by doubling all the entities that appear in Eqn. 1

$$O_4 + 2H_4 \to [H_8O_4] \to 2H_4O_2$$
1 vol. 2 vol. 2 vol.

A comparison of gas densities would give the same relative masses for the atoms of oxygen and hydrogen. Events have subsequently justified Avogadro's arbitrary choice of the smallest number, 2, for the number of atoms in the molecules of oxygen and hydrogen that satisfied the volume relationships. Ampère's choice of 4 seems to have been prompted by speculations concerning the structure of molecules in crystals in which he visualised molecules built from atoms placed at the corners of a regular tetrahedron.

Another set of molecular species that would have satisfied the observed volume relationships is tetra-atomic oxygen and diatomic hydrogen. Thus:

$$O_4 + 2H_2 \to [H_4O_4] \to 2H_2O_2$$

If this had been so, the ratio of the gas densities of oxygen and hydrogen would not have given the ratio of their atomic weights. There does not seem to be any evidence that during the first half of the 19th century, Avogadro's hypothesis was criticised and rejected because of these ambiguities. As it happened, Avogadro had intuitively hit on the now universally accepted interpretation of Gay-Lussac's Law and had obtained approximately correct values for the atomic weights of the common gaseous elements but all this was fully justified only after the lapse of some 50 years.

Despite the fact that it ultimately provided the key to the solution of the problem of an unequivocal system of atomic weights, Avogadro's hypothesis was almost completely ignored for 50 years, seldom if ever referred to and when it was, its origin was usually attributed to Ampère. The neglect of Avogadro is all the more difficult to understand when it is recalled that his two most important papers (those published in 1811 and 1814) appeared in a well-known French periodical—"*Journal de Physique de Chimie, de Histoire Naturelle et des Arts*". French chemists were so active about this time that chemistry was sometimes described as the "French Science".

In some ways it is surprising that Gay-Lussac failed to recognise the significance of Avogadro's work. There were circumstances connected with the publication of Avogadro's work which could have had some bearing on the question why it failed to receive the attention of chemists. The editor of the "*Journal de Physique*", Delametherie, was an ardent phlogistonist and thus an inveterate opponent of Lavoisier's views. It was because of this that Lavoisier participated in the founding, in 1789, of a new journal "*Annales de Chimie*" which has continued to appear regularly to this day. How long Delametherie persisted in his support of the phlogiston theory it is difficult to say but one thing is certain: "*Journal de Physique*" ceased publication in 1823. It is just possible that the impact of Avogadro's work was to some extent prejudiced by the nature of the journal in which he published it and by the fact it went out of existence so soon afterwards.

Avogadro does not appear to have discussed his work with scientists in other countries to any great extent either through visits or by correspondence. There is no evidence that he ever corresponded with Berzelius for example. The disturbed political conditions of his country in the early part of the nineteenth century may have had something to do with this.

Pauling[9] has suggested that Avogadro himself may have been largely responsible for the neglect of his ideas. "First, I think" he said, "that he could not imagine how great the value of his discovery was. We can see now in retrospect that almost the whole of the development of the science of chemistry has followed from the acceptance of correct atomic weights and the subsequent development of the chemical structure theory".

While Avogadro could hardly have been expected to foresee all the consequences of his work, he was not so unmindful of its possibilities as Pauling's analysis would suggest. Within 3 years of first putting forward his hypothesis, Avogadro[10] wrote:

"This hypothesis, once accepted, confirming in part the results to which Dalton, Davy and others have been led by appropriate deliberations on the masses of molecules [atoms] of different known substances based on their proportions in combination, furnishes us with a general method of correcting these results and thus perfecting the theory of definite proportions which is, or is going to be, the basis of all modern chemistry and the source of its future progress".

One of the greatest stumbling blocks in the way of accepting Avogadro's conceptual scheme was almost certainly the inability of chemists like Dalton and Berzelius to believe in the existence of polyatomic elementary molecules.

Dalton's concern over the seeming inconsistency of the gas densities of, say, oxygen and water vapor with "equal volume equal number of particles" hypothesis has already been referred to. He was equally unable to accept the idea of diatomic elementary molecules because of his belief that in the gaseous state, like atoms, each surrounded with an envelope of caloric, repelled one another. This assumption it will be recalled, was the basis of his first theory of the constitution of mixed gases and gaseous diffusion as it was earlier the basis of Newton's mathematical explanation of Boyle's Law. No one at that time, for that matter, for a long time afterwards, could suggest a reason for the stability of diatomic elementary molecules or why the process of combination of one atom with another of its kind should cease at 2; why not 4, 8 or more? That one atom should combine with another of its kind was inconsistent with the then widely accepted theory of chemical bond—the dualistic theory of chemical combination. According to this theory which was first proposed and staunchly supported by Ber-

References p. 83

zelius, perhaps the most influential chemist of the first half of the nineteenth century, atoms (and groups of atoms) entered into combination because of their supposedly opposite electrical polarity. If this were so, atoms of the same kind should repel, not combine with one another.

Toward the latter part of his scientific career, Avogadro turned his attention to liquids and solids. He became interested in atomic and molecular volumes. A few years earlier Kopp had calculated the atomic volume of elements by dividing the atomic weights (as given by Berzelius) by the corresponding densities of the solid or liquid substances. He found considerable differences from element to element. Avogadro did the same kind of calculations for compounds to obtain molecular volumes and found even greater variations from compound to compound. In order to reconcile experimental results with his theoretical views, Avogadro felt obliged to make what now seems to be a most astonishing assumption: that molecules of a gas may split into smaller molecules when the substance condenses to a liquid or solid. For example, he assumed that the molecule of gaseous chlorine split into four parts when it entered the liquid state. The conclusion that gas molecules may split into smaller molecules when gases condense to liquids must have done a great deal to shake the confidence of chemists in Avogadro's ideas.

It will be recalled that in his earlier papers, Avogadro did not use the term "atom". When, in later years he came round to doing so, he used it in a way that, to say the least, must have been confusing. In a paper dealing with specific heats of solids and liquids he spoke "of atoms of water" and of atoms being composed of "partial atoms".

In the same paper when discussing the transition of water to the liquid and solid state Avogadro[11] wrote:

"If we suppose that the atom of water in the liquid state was the same as that for water in the vapour state, representing the atoms by the density of the gases, it would be composed of one atom of hydrogen and half an atom of oxygen . . . but if we suppose that the atom of water divides into four parts in passing to the solid state, the atom of water in the solid state would then consist of one quarter of an atom of hydrogen and half an atom of oxygen".

Small wonder then that Avogadro's contemporaries were confused about the distinction between an atom and a molecule. Confusion of nomenclature and corresponding confusion of ideas may well have been the main cause in the delay in following up the consequences of Avogadro's work.

One may ask what then were Avogadro's main contributions to the atomic theory? First and foremost was his hypothesis, subsequently shown to be valid and the key to determining atomic weights; second was his demonstration that the hypothesis could be used as a basis for the comparison of molecular weights of gaseous substances simply by comparing gas densities; third was his deduction that the molecules of the common elementary gases are composite, consisting of two parts that we now call atoms.

REFERENCES

1 J.-L. GAY-LUSSAC, *Ann. Chim.*, 43 (1801) 367.
2 J.-L. GAY-LUSSAC, *Mém. Soc. d'Arceuil*, 2 (1809) 207. English translation, Alembic Club Reprint No. 4, *Foundations of the Molecular Theory*, Livingstone Ltd., Edinburgh, 1950.
3 M. L. CROSSLAND, The Origins of Gay-Lussac's Law of Combining Volumes, *Ann. Sci.*, 17 (1961) 1.
4 J. DALTON, *A New System of Chemical Philosophy*, Part 1, R. Bickerstaff, London, 1808, p. 559. Facsimile edition, W. Dawson and Sons Ltd., London.
5 A. AVOGADRO, Essay on a manner of determining the relative masses of the elementary molecules of bodies and the proportions in which they enter into these compounds. A translation of this paper: *J. de Phys.*, 73 (1811) 58. Alembic Club Reprint No. 4, Livingstone Ltd., Edinburgh, 1950.
6 J. C. GREGORY, *A Short History of Atomism*, Black, London, 1931, p. 109.
7 A. AVOGADRO, *Mem. R. Accad. Torino*, 26 (1821) 1–162.
8 A. M. AMPÈRE, *Ann. Chim.*, 89 (1814) 43.
9 L. PAULING, Amadeo Avogadro, *Science*, 124 (1956) 708.
10 A. AVOGADRO, Mémoire sur les Masses Relatives des Molécules des Corps Simples, *J. de Phys.*, 78 (1814) 131.
11 A. AVOGADRO, Mémoire sur les Chaleurs Spécifiques des Corps Solides et Liquides, *Mem. Soc. Ital. Sci. Res. Moderna*, 20 (1834) 80.

CHAPTER 6

The era of confusion and doubt

In the 50 years that followed Avogadro's statement of his all important hypothesis, development of the atomic theory was slow, halting and uncertain. The theory had its supporters but not a few of these abandoned it after initial enthusiasm. Some did so because of the failure to solve the problem of fixing true atomic weights. Others did so because they believed that the phenomena of chemistry could equally well be accounted for without assuming the existence of atoms—that atoms were purely imaginary entities. Nevertheless some progress was made in the understanding of different phenomena in terms of the theory even though the central problem of atomic weights remained unsolved.

Among the earliest supporters of the theory was Wollaston. Almost simultaneously he and Thomson discovered examples of the Law of Multiple proportions which they interpreted in terms of Dalton's ideas.

While studying oxalic acid and its compounds, Thomson[1] discovered "that there are two oxalates of strontian (strontium) the first obtained by saturating oxalic acid with strontian water (strontium hydroxide), the second by mixing together oxalate of ammonia and muriate of strontian (strontium chloride). It is remarkable that the first contains double the proportion of base contained in the second"*. This was one of the earliest examples of compounds exemplifying the Law of Mulple Proportions.

Very shortly after Thomson described his work on the oxalates of strontium, Wollaston[2] provided further evidence supporting the Law of Multiple Proportions in the form of the analyses of normal and acid sulphates and carbonates of potassium. He too was interested in the

---

* These are SrC_2O_4 and $Sr(HC_2O_4)_2$, respectively. In its most general form the Law states: If two substances A and B unite in more than one ratio, the various masses of A which combine with a fixed mass of B bear a simple ratio to one another. The Law is usually stated in terms of elements A and B.

salts of oxalic acid and described not only the neutral (1:1) and acid (1:2) salts of potassium but a third in which the amount of acid for reference quantity of base was 1:4 (the so-called quadroxalate). He tried hard to prepare a 1:3 salt but failed which puzzled him greatly. It caused him to write:

"When our views are sufficiently extended to enable us to reason with precision concerning the proportions of elementary atoms, we shall find the arithmetical relations alone will not be sufficient to explain their mutual action, and that we shall be obliged to acquire a geometrical conception of their relative arrangement in all three dimensions of solid extension".

Four years later, Wollaston[3] developed these spatial ideas in a truly remarkable way in his Bakerian Lecture to the Royal Society. Like Hooke before him, he tried to account for the forms of crystals by the packing of spherical particles but he far exceeded Hooke in the elaboration of his ideas. Hooke had shown how the close-packing of equal spheres could be used to explain the development of the various crystal faces of alum. In suggesting that metals "behave as if they were composed of spherical particles", Wollaston came very close to the modern view of their structure and was thoroughly well acquainted with the geometry of close packed structures of equal spheres.

"Let a mass of matter be supposed to consist of spherical particles, all of the same size but of two different kinds in equal numbers represented by black balls and white balls; and let it be required that in their perfect intermixture every black ball shall be equidistant from all surrounding white balls and that all adjacent balls of the same denomination shall be equidistant from each other. I say then that these conditions will be fulfilled if the arrangement be cubical". This was intended as a description of the probable structure of boracite but it was quite wide of the mark—the actual structure being much more complex. However, the description more nearly fits the known structure of potassium chloride which is a face-centered cubic ionic crystal.

Wollaston concluded "any attempt to trace a general correspondence between the crystallographical and supposed chemical elements of bodies must, in the present state of these sciences, be premature".

Nevertheless, as the accompanying diagrams (Fig. 5) indicate, he showed remarkable insight and anticipated by a century some of the

References pp. 102-103

kinds of structures revealed by X-ray crystallography. Only after the discovery of X-rays and their diffraction by crystals was it possible to discover the sizes, shapes and arrangement of atoms in crystals.

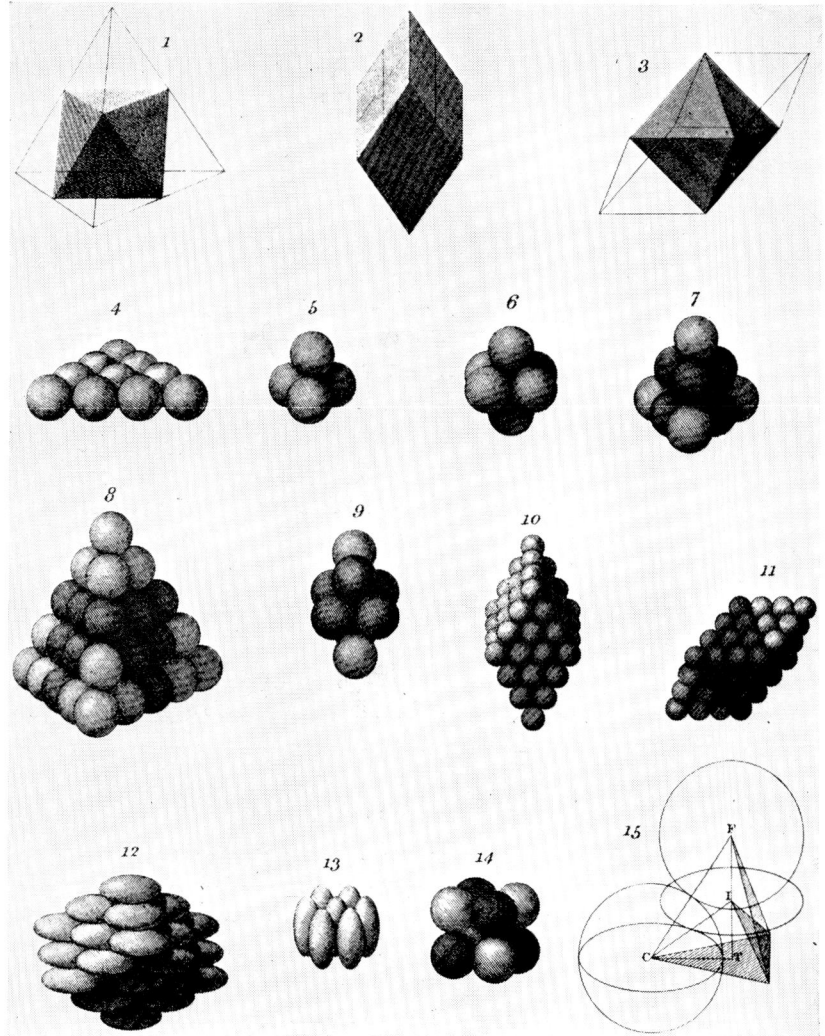

Fig. 5. The arrangements of particles in crystals suggested by Wollaston (1812).

Extraordinary as it may now seem, Wollaston after this remarkable flight of imaginative insight abandoned the atomic hypothesis to the point of deciding that it was fruitless to try to discover the relative weights of atoms. He was apparently unaware of the work of Avogadro. The difficulty of finding the number of atoms in a molecule seemed to him insuperable and he finally decided that chemistry was best served by using the results of direct chemical analysis—namely, chemical equivalents or combining proportions. Dalton's atomic weights ($O = 8$, $N = 5$, $C = 4.5$) were, in fact, chemical equivalents or rough approximations thereto.

Since, in the next 50 years or so there was to be great confusion over atomic and equivalent weights, one being taken for the other, it will be worth explaining the difference between the two in some detail. The chemical equivalent of an element is defined in terms of some standard, for example, the amount of the element which combines with eight grams of oxygen. Equivalent weights are the direct results of chemical analysis and their measurement is quite independent of the atomic theory. Many elements possess more than one equivalent weight. The atomic weight of an element may be identical with its equivalent weight or it may be related to it by some small factor—two, three, four or five. The atomic weight of oxygen is double its equivalent weight, nitrogen three times, carbon four, phosphorus five (or three) times its atomic weight.

The essential problem of fixing atomic weights was to discover this factor. Wollaston decided it could not be done.

The greatest chemist of the early 19th century to accept the atomic theory was undoubtedly Berzelius (1779–1848)[4]. The son of a schoolmaster, he was born at Wafnersunda, Sweden. As a student at secondary school, Berzelius showed no great promise; his leaving certificate bore the statement that he "justified only doubtful hopes". He went on to study medicine at Uppsala where he graduated after submitting a dissertation on the effect of galvanic currents on various chronic diseases—rather an unlikely subject for a man who was to become a most outstanding chemist. He accepted Dalton's theory with considerable enthusiasm and was especially impressed by the light it shed on the law of multiple proportions. "Without Dalton's theory" he said "this law would be a mystery". "If Dalton's hypothesis of Combination in Multiple Proportions is found to be true, it will throw such new light on the theory of chemical affinity, that it will be the greatest advance towards the perfection of chemistry as a science, that has yet been

References pp. 102-103

made". Berzelius himself established with greater accuracy than any previous investigator, the existence of many instances of this law.

He became intensely interested in the subject of atomic weights and devoted a great deal of his working chemical life to their accurate measurement.

Jöns Jacob Berzelius

"I perceived" Berzelius[5] said, "that if the light that was now to be shed upon the whole of science was to be increased, the atomic weights of as large a number of elements as possible and especially those of commonly occurring elements, must be determined with the greatest accuracy possible". Over the period 1808–1818, Berzelius determined single-handed the atomic weights of almost all the elements known at the time. From these values and his analytical results, he assigned formulae for over 2,000 compounds which, it must be admitted, was a truly remarkable achievement.

Berzelius was greatly impressed with Gay-Lussac's Law of combining volumes and made use of it in determining molecular formulae. To do this he assumed that equal volumes of *elementary* gases under the same conditions of temperature and pressure contained the same number of atoms. Thus for the reaction between oxygen and hydrogen he would have written something like the following;

$$\text{O} + 2\text{H} \rightarrow \text{H}_2\text{O}$$
$$\text{1 vol.} \quad \text{2 vol.} \quad \text{2 vol.}$$
$$n \text{ atom} \quad 2n \text{ atoms} \quad n \text{ molecules}$$
$$\left(\frac{n}{2} \text{ molecules per unit volume}\right)$$

This interpretation led Berzelius to the correct relative weights of the atoms of hydrogen and oxygen since both gases are diatomic and, accordingly, he wrote the formula for water H_2O.

Early in his career Berzelius[6], revived and extended the use of letter symbols for the elements which he did, curiously enough, first in an English journal.

Dalton's symbols, though they had the advantage of encouraging chemists to visualise atoms and their arrangement in molecules, were far too cumbrous and unsuited for printing. "The chemical signs" said Berzelius "ought to be letters, for the greater facility of writing and not to disfigure a printed book. I shall take therefore for the chemical sign the initial letter of the Latin name of each elementary substance". Where there were two names with the same initial letter, he used the first and second letters for one of them.

The history of chemical symbolism is an interesting one but it would take us too far afield to follow it in all its details[7]. Berzelius first wrote the formula for the molecule of water $2H+O$; later, it became H^2O. At one time, when he tried to abbreviate formulae by using a special symbol for oxygen because of its seemingly central position in the hierarchy of elements, he wrote a dot above the symbol of the element with which it combined. Thus, water became $\dot{H}^2$. A further modification adopted by Berzelius was the use of barred symbols to denote two atoms $-H-$, a device which for a time was to cause great confusion. Fortunately for chemistry, the dot and bar conventions were both dropped. Superscripts became subscripts[7], and water became H_2O.

Berzelius pursued the study of atomic weights with great vigour and

References pp. 102-103

published revisions of his 1814 table in 1818 and again in 1826. These are shown in Table 3*.

TABLE 3

THE ATOMIC WEIGHTS OF BERZELIUS

Element	1814	1818	1826	Modern
O	16	16	16	16
S	32.16	32.19	32.19	32.07
P	26.80	32.88×2	31.38×2	30.98
M	22.33	22.82	—	—
Cl	—	—	35.41	35.46
C	11.99	12.05	12.23	12.01
Nt	12.73	12.36	—	—
N	—	14.18×2	14.16	14.008
H	1.062	0.995	0.998	1.008
As	67.19×2	75.26×2	75.21	74.92
Cr	56.64	56.29	56.29	52.01
Sb	129.0×2	129.0×2	129.03	121.76
Si	32.46×$\frac{3}{2}$	31.62×$\frac{3}{2}$	29.58×$\frac{3}{2}$	28.09
Hg	202.5×2	202.5×2	202.5	200.61
Ag	107.52×4	108.12×4	108.12×2	107.88
Cu	64.51×2	63.31×2	66.31	63.54
Bi	189.2×$\frac{3}{2}$	189.2×$\frac{3}{2}$	212.80	208.99
Pb	207.76×2	207.12×2	207.12	207.21
Sn	117.6×2	117.6×2	117.6×2	118.70
Fe	55.5×2	54.27×2	54.27	55.75
Zn	64.52×2	64.52×2	64.51	65.38
Mn	56.92×2	56.92×2	56.92	54.94
Al	27.44×2	27.38×2	27.38	26.98
Mg	25.23×2	25.33×2	25.33	24.32
Ca	40.81×2	40.96×2	40.96	40.08
Ba	136.73×2	137.1×2	137.1	137.36
Na	23.17×4	23.27×4	23.27×2	22.99
K	39.12×4	39.19×4	29.19×2	39.10

His values were based on O = 100 as standard. In order to see at a glance how they compare with present day values, all his results shown in Table 1 are recalculated to O = 16. When it came to fixing the atomic weights of the metals, Berzelius was in precisely the same position as Dalton and Avogadro. He relied on a set of arbitrary rules as did Dalton.

* This table is taken, with minor modifications, from J. R. PARTINGTON, *History of Chemistry*, Vol. IV, p. 166.

The combinations of elements A and B could be

(1) A+B, 2B, 3B . . . . the limit being 12B.
(2) 2A+3B or 5B, 7B. It was not possible to distinguish between A+B and 2A+2B.

It will be seen that by 1826 his values were in general remarkably close to the modern values with about four notable exceptions Ag, Si, Na and K.

Berzelius' values for hydrogen, oxygen, nitrogen and chlorine agreed with those of Avogadro though as already pointed out he did not make use of Avogadro's hypothesis. The atomic weight of carbon had been determined correctly on the assumption that carbon monoxide was CO and carbon dioxide was CO_2; having fixed the atomic weight of oxygen ($O = 16$), both Avogadro and Berzelius calculated correctly the atomic weight of carbon which was consistent with the gas densities of carbon monoxide and carbon dioxide. Berzelius' value for the atomic weight of sulphur (32 in all three tables) was calculated from the gas density of sulphuretted hydrogen on the assumption that its formula is analogous to that of water. He also used the analogy between sulphides and oxides, *i.e.* CuS and CuO. The value for sulphur is the modern one. In view of later developments the value for silver is interesting. He found that 13.516 g of silver combined with 1 g of oxygen to form silver oxide. In order to calculate the atomic weight of silver from this result, Berzelius had to choose between the possible formulae Ag_2O, AgO and AgO_2.

If the first was the correct one, the atomic weight of silver was 108.13; if the second 216.16; the third 423.5. His first choice (in the 1814 and 1818 tables) was AgO_2 and he therefore listed the atomic weight of silver ($O = 16$) as 432.52, four times the accepted value. He assumed an analogous formula for the oxide of platinum IV and as a result obtained the value of 193 for the atomic weight of platinum (approximately the modern value, 195.1). The atomic weights of sulphur and platinum have been mentioned here because of their bearing on the next important development in the history of the atomic theory. No one worked so skilfully and indefatigably as did Berzelius towards the objective of measuring the atomic weights of all the known elements. "He was a fine craftsman and we owe to him much of our laboratory technique". He displayed remarkable chemical insight and versatility

References pp. 102-103

of approach in his attack on the problem and the great majority of his values were approximately the modern ones. By using the "equal volume equal number of atoms" assumption, he deduced correctly the atomic weights of hydrogen and nitrogen and wrote the formula for ammonia as NH_3 (1818). On the strength of the assumed analogy between ammonia, phosphine (PH_3) and arsine (AsH_3), he deduced nearly correct values for the atomic weights of phosphorus and arsenic.

He made great use of Mitscherlich's principle[8] of isomorphism—"the same number of atoms combined in the same manner produces the same crystalline form; crystalline form is independent of the chemical nature of the atoms and is determined solely by their number and mode of combination (see also ref. 9). As evidence in support of the atomic theory Berzelius regarded the discovery of isomorphism as "the most important since the establishment of the doctrine of chemical proportions" and made use of it in his atomic weight determinations. For example, Mitscherlich found that potassium permanganate and potassium perchlorate were isomorphous. The three dimensional arrangement of atoms in a crystal of potassium perchlorate is identical with that of potassium permanganate, except that every manganese atom is replaced by a chlorine atom. In these compounds, 55 parts by weight of manganese replace 35.5 parts per weight of chlorine. Since 35.5 is the atomic weight of chlorine, that of manganese must be 55.

Similarly, potassium manganate is isomorphous with potassium chromate, sulphate and selenate. In this series, 55 parts by weight of manganese are replaced by 32 of sulphur, 79 of selenium and 52 of chromium. Since 55 is the atomic weight of manganese, the atomic weights of the other elements are sulphur 32, selenium 79 and chromium 52. Discovery of the principle of isomorphism not only provided an aid in the determination of atomic weights, it also supported the atomic theory insofar as the theory provided a simple explanation or interpretation of the principle. The atomic theory stimulated a search for further examples of isomorphism.

Berzelius admitted the great difficulty of achieving the task he had set himself when he said "to hit upon what is true is a matter of luck, the full value of which is, however, only realised when we can prove what we have found is true. Unfortunately, in these matters, the certainty of our knowledge is at so low a level that all we can do is to follow the lines of greatest probability" (*Jahresbericht* 1828).

The task of determining atomic weights could be roughly divided into two parts: firstly the determination of the atomic weights of volatile elements and those elements that form volatile compounds; secondly, the determination of the atomic weights of the metals of which, at that time, very few were known to form stable volatile compounds. For the former, the application of Avogadro's hypothesis provided the answer. Mitscherlich's discovery of isomorphism provided part of the answer for the second, but another discovery was to pave the way for a more general answer.

While Berzelius was a visitor at the home of the Society of Arceuil in Paris, he collaborated with a young chemist, Dulong, whom he described as having the most brilliant mind in the world of chemistry at that time. In the spring of 1819, they determined with considerable accuracy the atomic weight of hydrogen (on the basis of $O = 100$) from the O/H ratio in water by reducing copper oxide with pure hydrogen gas, weighing the water formed and determining the change in weight of copper oxide.

Dulong's originality was, however, more clearly evident in another investigation undertaken about this time. Dulong and Petit[10] set out to study the relationship between the specific heats of the elements and the heat capacity of atoms.

"Convinced that certain properties of matter would present themselves in simpler forms and that they could be expressed by more regular and less complicated laws if it were possible to relate them to the elements on which they directly depend, we have attempted to introduce the best established results of the atomic theory into the study of some of the properties that seem to be most intimately connected with the individual action of the material molecules. Among such properties we have selected for our special attention those which depend upon the action of heat". (See also refs. 11 and 12.) In particular, they chose to study specific heats* of elements. Realising that a

---

* Specific heat of a substance is the ratio between the amount of heat required to raise any weight of the substance through any temperature range and the amount of heat required to produce the same effect on the same weight of an arbitrarily chosen standard.

$$\text{Specific heat} = \frac{\text{Heat capacity of A}}{\text{Heat capacity of standard}}$$

Standard substance is water—mass in g, temperature range 1°C.

comparison of specific heats alone was insufficient to throw light on the heat capacity of atoms, they decided to compare the heat capacities of equal numbers of atoms.

The atomic weight in grams of different elements will contain the same number of atoms—this they realized quite clearly. Thus, to find the heat capacity of different atoms, they multiplied specific heat by atomic weight. The product, they found, was approximately constant.

It is worth while to examine the original table published by Dulong and Petit (Table 4 A). The specific heats were the results of their own measurements. Strangely enough, they do not state the origin of the atomic weights they used though it is evident they have been calculated on the base $O = 1$. It is practically certain that, in the main, they used the values published by Berzelius in 1818. In Table 4B the atomic weights have been calculated to the base $O = 16$. The last two columns of the table show the atomic weights adopted by Berzelius about this time and the factor by which they must be multiplied to give the values used by Dulong and Petit.

On their results they commented as follows: "Simple inspection of these numbers reveals an agreement too marked in its simplicity not to show at once the existence of a physical law susceptible of being generalised and extended to all elementary substances. The products which express the heat capacities of the different elements approach so nearly to equality that it is impossible for the very slight differences not to be due to unavoidable errors, either in the determination of the capacities, or in the chemical analyses (*i.e.* the atomic weight). The number and diversity of the substances with which we have worked do not permit the relation we have just indicated to be mere chance and hence it is proper to deduce from this relation the following law: All atoms of simple substances have the same capacity for heat". This statement would seem to indicate they had the utmost confidence in the values of the atomic weights they chose. Why did they modify the values of Berzelius in the way they did?

To understand the situation, let us assume for the moment that all the atomic weights published by Berzelius in 1818 were correct and that the product of specific heat and atomic weight is in fact, constant. Dulong and Petit would have obtained a constant product even if they had halved *all* the correct atomic weights.

The extraordinary thing is that they chose to halve Berzelius' values

TABLE 4A

ATOMIC HEATS: ORIGINAL TABLE OF DULONG AND PETIT (1819)

CHALEURS SPÉCIFIQUES (1).		POIDS RELATIFS des atomes (2).	PRODUITS du poids de chaque atome par la capacité correspondante.
Bismuth,	0,0288	13,30	0,3830
Plomb,	0,0293	12,95	0,3794
Or,	0,0298	12,43	0,3704
Platine,	0,0314	11,16	0,3740
Etain,	0,0514	7,35	0,3779
Argent,	0,0557	6,75	0,3759
Zinc,	0,0927	4,03	0,3736
Tellure,	0,0912	4,03	0,3675
Cuivre,	0,0949	3,957	0,3755
Nickel,	0,1035	3,69	0,3819
Fer,	0,1100	3,392	0,3731
Cobalt,	0,1498	2,46	0,3685
Soufre,	0,1880	2,011	0,3780

TABLE 4B

ATOMIC HEATS (DULONG AND PETIT) RECALCULATED

Element	Specific Heat	Atomic Weight	Atomic Heat	Atomic Weight Berzelius 1818	Factor for Conversion
Bismuth	0.0288	212.8	6.128	283.8	$\frac{3}{4}$
Lead	0.0293	207.2	6.070	414.2	$\frac{1}{2}$
Gold	0.0298	198.9	5.926	397.8	$\frac{1}{2}$
Platinum	0.0314	178.6	5.984	178.4	1
Tin	0.0514	117.6	6.046	235.2	$\frac{1}{2}$
Silver	0.0557	108.0	6.014	432.5	$\frac{1}{4}$
Zinc	0.0927	64.5	5.978	129.0	$\frac{1}{2}$
Tellurium	0.0912	64.5	5.880	129.0	$\frac{1}{2}$
Copper	0.0949	63.31	6.008	126.6	$\frac{1}{2}$
Nickel	0.1035	59.0	6.110	118.3	$\frac{1}{2}$
Iron	0.1100	54.27	5.970	108.55	$\frac{1}{2}$
Cobalt	0.1498	39.36	5.896	118.3	$\frac{1}{3}$
Sulfur	0.1880	32.19	6.048	32.19	1

References pp. 102-103

for seven of the metals. The only values of Berzelius they used unchanged were those for platinum and sulphur. The product of specific heat and atomic weight for these two elements is half that of the seven elements just referred to (if the atomic weights of Berzelius are used for the latter).

It is difficult to avoid the conclusion that Dulong and Petit assumed that the product of specific heat and atomic weight is constant and that they decided the appropriate thing to do was to bring all their results into line by so adjusting the atomic weights that the product was equal to the minimum value—that obtained for platinum and sulphur. By this procedure they made the atomic weights of the seven metals approximately the same as their present day values. That Berzelius obtained a value of the atomic weight of platinum approximately* the modern one was the result of an arbitrary assumption that platinum formed an oxide $Pt+2O$ similar to that of silver $Ag+2O$. This, of course, gave a value for silver four times the correct value but the correct value for platinum.

On the other hand, Berzelius based the value for sulphur (32) which is close to the modern one on the soundly based analogy between sulphur and oxygen.

A remarkable feature of Dulong and Petit's work is that they should have chosen a value for the product of atomic weight and specific heat that brought the atomic weights of the metals they studied close to the modern values. Newton is reported to have said "no great discovery is made without a bold guess". If ever a discovery fitted this description it was that of Dulong and Petit.

About 20 years later, Regnault[13] vindicated and greatly extended and strengthened the basis of Dulong and Petit's law. Further work, however, was to show that the law, as stated, did not apply to elements like carbon, silicon and boron. For all elements, atomic heat is a function of temperature; for metals, it approaches a limiting value of 6 calories/degree at room temperature; on the other hand, the atomic

---

* The value for platinum in the 1814 table of Berzelius was 12.06 ($O = 1$). By 1818 he had changed this to 12.15. Dulong and Petit used a much less accurate value 11.16; and, at the same time, made an arithmetical error in calculating the product 11.16×0.0314—which is 0.350 not 0.3740 as quoted in their table. The origin of the value 11.16 for the atomic weight of platinum is uncertain. Their value for the specific heat of platinum 0.0314 is in error. The modern values are 195.1 and 0.0324, respectively.

heat of carbon at room temperature is about 1.35; at 968° C, it is still only 5.546 (ref. 14). A theoretical basis in terms of the kinetic theory and quantum theory has since been found for the law. Dulong and Petit's law provided support for the atomic theory in an indirect kind of way; here was a measurable physical quantity, namely atomic heat, which was directly proportional to the number of atoms studied whatever their kind. The law was consistent with and explicable in terms of the atomic theory. Its importance lay in the fact that it provided a means for determining approximate values of the atomic weights of those elements to which Avogadro's method was not applicable. These were mainly the metals, for which few or no stable, volatile compounds suitable for vapor density measurements were then available. The following data will serve to illustrate the use of the law for finding approximate values of atomic weights:

TABLE 5

ATOMIC WEIGHTS DEDUCED FROM SPECIFIC HEATS

Element	Equivalent weight	Spec. heat	Approx. at wt. 6.4/Spec. heat	Accepted at. wt.
Sodium	23.06	0.2934	21.8	23.06
Potassium	39.15	0.1655	38.8	39.1
Lithium	7.03	0.9408	6.8	7.03
Calcium	20.0	0.1732	37.6	40.0
Magnesium	12.18	0.2499	25.6	24.46

The equivalent weight, now defined as that weight of an element which combines with 8.00 g of oxygen may be determined with a high degree of accuracy by chemical analysis. Application of Dulong and Petit's law provides the appropriate factor by which the equivalent weight of an element must be multiplied in order to obtain the atomic weight. For calcium and magnesium this factor is, to the nearest whole number, 2. Since the atomic weight derived from the specific heat is only an approximate value, the factor calculated from the ratio of the atomic weight to the equivalent is a correspondingly approximate value. Where it is possible to measure both the atomic and equivalent weight accurately, the ratio is very close to a small whole number. The significance of this number will be discussed in Chapter 9.

References pp. 102-103

What influence did the work of Dulong and Petit have on Berzelius? An inspection of the 1826 column of Table 3 will show that Berzelius by halving his values for the atomic weights of copper, iron and zinc, made them close to the values accepted today. In view of his very high opinion of Dulong's ability as a chemist it is extraordinary that Berzelius did not correct the atomic weight of silver as he could have done. His value was 432.52; whereas as the specific heat clearly indicated that it was approximately $\frac{1}{4}$ of this value. He did, in fact, halve his original value for silver and he may have felt that this was going far enough. His failure to correct the value for silver was later to cause considerable confusion among organic chemists who sometimes used silver salts in the determination of the equivalent weight of organic acids.

About this time, Dumas (1800–1884), one of the most distinguished chemists of the 19th century, set out to establish an unequivocal system of atomic weights. In a paper entitled[15] "*Sur quelques points de la théorie atomique*", he declared:

"The object of these researches is to replace by definite conceptions the arbitrary data on which the whole of the atomic theory is based".

Dumas set out on his ambitious task along sound lines: he devised a new method for measuring vapor densities at temperatures above room temperature and attempted to apply Avogadro's hypothesis. His method of measurement was a very simple one. It consisted of filling a glass bulb of known weight and volume with some of the substance being studied and then placing it in a bath of liquid whose boiling point is well above that of the substance being investigated. When all the substance has been boiled away and the bulb is full of its vapour, the neck is then sealed and after cooling is weighed. From a knowledge of the volume of the flask and the weight of the vapour it contained at the temperature of the bath and atmospheric pressure, the vapour density may be calculated. This technique was a valuable addition to experimental chemistry and enabled the accumulation of data that was to prove essential for the later development of the atomic theory.

Dumas applied his method to the relatively volatile elements, mercury, sulphur, phosphorus and arsenic with results that were to him and many of his contemporaries, extremely puzzling. The density of mercury vapour proved to be about one hundred times that of hydrogen at the same temperature and pressure. On the basis of $H_2 = 2$, this was interpreted by Dumas to mean that the atomic weight of mercury is 100; whereas the application of Dulong and Petit's law indicated that it should be about double this value. Dumas had assumed that the molecule of mercury was, like those of the common gases, diatomic. His results for sulphur were, if anything, even more puzzling.

Dulong and Petit's rule indicated that the atomic weight of sulphur was 32 but vapour density measurements indicated a value of 96 (if the molecule was S_2). Similarly, vapour density measurements on phosphorus and arsenic indicated atomic weights about double the values to be anticipated on the assumption that the molecules present in the gas phase were P_2 and As_2.

Although Dumas stated Avogadro's hypothesis correctly in his 1826 paper, that is, in terms of molecules, 2 years later he stated that equal volumes of different gases under the same conditions of pressure and temperature contained the same number of *atoms*. In his textbook[16] he spoke of the atoms of a gas always containing a certain number of molecules. In other words he reversed the usage of the terms atom and molecule.

Quite apart from this confusion in nomenclature there were inherent difficulties in the way of interpreting the significance of gas density measurements. Measurements of gas densities of the elements alone provided no clue to their complexity in the gaseous state. Except in rare instances, atoms do not occur in nature as isolated entities; some notable exceptions to this statement are mercury vapour and the noble gases of the atmosphere—helium, neon, argon, krypton and xenon. In all the last-named gases the particles consist of single atoms. If all the elementary gases and vapours had been monatomic the measurement of atomic weights by a comparison of gas densities would have been straightforward. As it is, the molecules of some elements consist of two atoms, others four and occasionally eight. There were at that time no chemical or physical reasons for deciding which degree of molecular complexity would be exhibited by any particular element. Hence Dumas' difficulty in interpreting the results of his vapour density measurements.

Dumas[17] was still confused about atoms and molecules in 1832 when he wrote:

"Though it is quite easy to establish the ratio in which elements

References pp. 102-103

combine, it is very difficult to estimate the actual number of atoms which enter into these combinations. Berzelius in his treatise on chemical proportions, which marks so important an epoch in the history of science ... was the first to attack this problem in its full scope. Without any rules to guide him, he fixed by intuition the atomic weight of each substance, and usually allowed himself to be influenced by analogies which subsequent experience has only tended to confirm. But chemists have always wished that this arbitrary method, so successfully used by Berzelius, might be supplanted by something more fixed, more susceptible to all kinds of intellect, and less subject to the capricious modifications of each writer".

Dumas was unable to find any method "more fixed" and less arbitrary. A few years later he was to write somewhat despairingly[18]:

"It is my belief that the equivalents of chemists like Wenzel and Mitscherlich which we call atoms are nothing other than molecular groupings. If I were master, I would abolish the word atom from science, persuaded as I am that it goes beyond experience; always in chemistry we ought not to go beyond experience".

Dumas abandoned the atomic theory and adopted the positivistic philosophy which rejects the explanation of phenomena in terms of unobserved entities. In adopting this standpoint Dumas was almost certainly influenced by the philosopher Comte[19] who had expressed similar views about the atomic theory: "The real mode of agglomeration of elementary particles is, and ever must be unknown to us, and therefore no proper object of our study".

An explanation of Dumas' results was given by one of his students, Gaudin[20], in a paper published in 1833. He made use of atomic weights for these elements published by Berzelius ($Hg = 202.5$; $S = 32.19$; $P = 31.4$; $As = 75.3$) all of which happen to be approximately the values accepted today. Dividing the molecular weights of these elements determined from vapor density measurements by the corresponding atomic weights Gaudin deduced that mercury was monatomic (Hg), sulphur hexatomic (S_6)* and phosphorus and arsenic were tetraatomic (P_4 and As_4).

That his explanation was the correct one only became certain after

* Modern measurements have shown that sulphur vapor is a complex mixture of molecular species (S_8, S_6, S_4 etc.) the proportions of which depend on the temperature of the vapor.

a universally acceptable method of fixing atomic weights had been established.

Gaudin clearly understood the distinction between atoms and molecules. This is evident from his "volume diagrams" in which he represented the relationships between the combining volumes of different gases. Notice his use of Daltonian symbols for atoms and molecules.

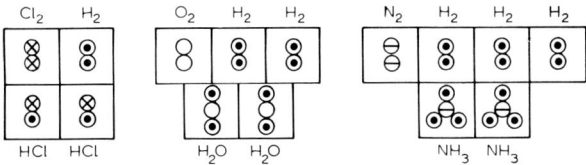

Fig. 6. Gaudin's gas volume diagrams (1833).

Putting the distinction into words he said "An atom will be for us a small spherical homogeneous particle, essentially indivisible, whilst a molecule will be an isolated group of atoms of whatever number and whatever kind".

Gaudin is notable for being one of the very rare individuals, in the period between 1811 and 1858, who understood the application of Avogadro's hypothesis to the combination of gases. Ironically enough he attributed the hypothesis to Ampère making no mention whatever of Avogadro.

Partington[21] says that "Gaudin was outside the academic clique in Paris". Consequently, it is not altogether surprising that little notice appears to have been taken of his work. The little notice of it taken outside Paris was in at least one instance unfavourable: Berzelius commented on it adversely. And so it was that for some time to come this clear statement of Avogadro's interpretation of the reactions between gases exerted very little influence on the development of the atomic theory.

An examination of Berzelius' 1826 table of atomic weights (see Table 1) will show that with the exception of the alkali metals, silver and silicon, the atomic weights are remarkably close to the modern values. Yet by 1840, atomic weights so laboriously and accurately measured by Berzelius were widely distrusted and often abandoned altogether in favour of equivalents. Despite his remarkable experimental achievement, Berzelius had failed to find an unequivocal, universally accept-

References pp. 102–103

able method for measuring the true relative weights of atoms. Added to this was the fact that the dualistic theory of chemical combination had lost ground and as a result Berzelius' influence on chemistry generally was on the wane. Doubts caused by the vapor density measurements of Dumas still lingered on and the atomic theory entered a period of widespread scepticism. The positivistic outlook gained ground among chemists causing them to reject atoms as imaginary and nonexistent entities. Well known chemists like Liebig and Gay-Lussac despaired of ever being able to measure with certainty the true relative weights of atoms.

Chemists began to urge the general adoption of equivalent weights as the safest guide. "The study of chemistry would be made infinitely easier" said Liebig[22] "if all chemists agreed to return to equivalents".

L. Gmelin, author of the famous handbooks of inorganic chemistry, was a prominent advocate of adoption of combining weights such as $O = 8$, $Ca = 20$.

Thus, despite the great advances made in the first two decades after Dalton's first statement of the chemical theory of atoms and the high hopes of its supporters, Avogadro, Thomson, Berzelius and Dulong, progress, by the middle of the century, had almost ceased.

The undercurrent of scepticism about the atomic theory that began with Davy, Wollaston and Dumas, now reached its greatest strength and continued but with diminishing strength for the next 20 years or so.

REFERENCES

1 T. Thomson, *Phil. Trans.*, 98 (1808) 63. See also Alembic Club Reprint No. 2, Foundations of the Atomic Theory, Livingstone Ltd., Edinburgh, 1948.
2 W. H. Wollaston, *Phil. Trans.*, 98 (1808) 96.
3 W. H. Wollaston, *Phil. Trans.*, 103 (1813) 51.
4 J. E. Jorpes, *Jacob Berzelius, His Life and Work*, translated from the Swedish by Barbara Steele, Almquist and Wiksell, Stockholm, 1966.
5 H. Hartley, The Place of Berzelius in the History of Chemistry, *K. Svenska Veten. Årsbok*, Stockholm, 1948.
6 J. J. Berzelius, *Ann. Phil.*, 3 (1814) 51.
7 M. P. Crossland, *Historical Studies in the Language of Chemistry*, Heinemann, London, 1962, p. 272.
8 E. Mitscherlich, *Ann. Chim. Phys.*, 19 (1821) 350.
9 S. I. Morrow, One Hundred Years of Isomorphism, *J. Chem. Educ.*, 46 (1969) 580.

10 P. L. Dulong and A. T. Petit, *Ann. Chim.*, 10 (1819) 395.
11 P. Lemay and R. E. Oesper, Pierre Louis Dulong, His Life and Work, *Chymia*, 1 (1948) 171.
12 R. Fox, Background to the Discovery of Dulong and Petit's Law, *Brit. J. Hist. Sci.*, 4 (1968) 1.
13 V. Regnault, *Ann. Chim. Phys.*, 73 (1840) 1.
14 F. Seitz, *Modern Theory of Solids*, McGraw-Hill, New York, 1940.
15 J. B. Dumas, *Ann. Chim. Phys.*, 33 (1826) 337.
16 J. B. Dumas, *Traité de Chimie Appliqué aux Arts*, Vol. 1, Paris, 1828, p. 40.
17 J. B. Dumas, *Ann. Chim. Phys.*, 50 (1832) 170.
18 J. B. Dumas, *Leçons sur la Philosophie Chimique*, Paris, 1836, p. 290.
19 I. A. M. Comte, *Course de Philosophie Positive*, Paris, 1832.
20 A. M. Gaudin, *Ann. Chim.*, 52 (1833) 113.
21 J. R. Partington, *A History of Chemistry*, Macmillan and Co., London, Vol. 4, 1964, p. 11.
22 J. von Liebig, *Ann. Pharm.*, 31 (1839) 35.

CHAPTER 7

The successful solution of the problem of determining atomic weights

"If you are in the very thickest fog you are not confused because you see nothing to confuse you. But as the fog gradually disperses you catch sight indistinctly of a bit of something here and other bit of something there; you are not quite sure what each is or how it fits in with the rest. Then you are confused until it is clear enough to see everything".

This is an apt description[1] of the progress of efforts to determine the relative weights of atoms. By the middle of the 19th century the confusion was at its height. Another 20 years were to elapse before it cleared.

Although Berzelius used correct values for the atomic weights of carbon (12), hydrogen (1), oxygen (16) and nitrogen (14), his formulae for organic acids were double the true ones. This stemmed from the fact that he used their silver salts to calculate molecular weights which he defined as the weight that combines with one molecule of silver oxide. Since he assumed the formula AgO (Ag = 2×108) for the oxide Ag_2O, his molecular weights and formulae were double the true value. Thus, for acetic acid he wrote $C_4H_8O_4$ instead of $C_2H_4O_2$. The irony of this situation is that Berzelius who worked so indefatigably and successfully to determine atomic weights, should have failed to apply Dulong and Petit's law to silver which would have given him correct value for the atomic weight of silver. He applied the law to correct his earlier values for the atomic weights of other metals; why he failed to do so for silver is a mystery.

Chemists were perplexed by worse troubles than doubling of formulae.

Kekulé[2] (1829–1896) recorded that 19 different formulae for acetic acid could be found in the literature. Five of them are shown below:

(1) $C_4H_4O_4$ (C = 6, H = 1, O = 8)
(2) $C_8H_4O_4$ (C = 3, H = 1, O = 8)

(3) $C_4H_8O_4$ (C = 12, H = 1, O = 16)
(4) $C_2H_4O_2$ (C = 12, H = 1, O = 16)
(5) $C_2H_4O_4$ (C = 12, H = 1, O = 8)

It was almost as if every chemist had his own system of atomic weights. With such confusion about even the empirical formulae of substances, it is not to be wondered at that progress in the understanding of valency was slow.

Gradually the mists of confusion began to clear. Prominent contributors facilitating the process during the 1850's were Gerhardt and Laurent. Gerhardt[3] (1816–1856) noticed that, if in what he called double decomposition reactions, the formulae of organic compounds were written in the way frequently adopted (*i.e.* doubled), then water, carbon dioxide, hydrochloric and ammonia were eliminated as double molecules, *e.g.* H_4O_2 as in:

$$C_{24}H_{10}H_2 + H_2O_2N_2O_4 = H_4O_2 + C_{24}H_{10}N_2O_4$$
$$C_{14}H_{12}O_6 = C_2O_4 + C_{12}H_{12}O_2$$

In modern form:

$$C_6H_5H + HNO_3 = H_2O + C_6H_5NO_2$$
$$C_7H_6O_3 = CO_2 + C_6H_6O$$

From this, he concluded that if Berzelius' formulae for the inorganic molecules CO_2 and H_2O were correct, then the formulae for organic compounds should be halved. He reinforced this belief by reducing formulae of volatile organic compounds to what he called the "two volume" basis, that is by measuring by the Dumas method, the weight of the volatile compound that occupied the same volume as 2 g of hydrogen under the same conditions of temperature and pressure. The formula so obtained was half that given by Berzelius. In doing this, however, Gerhardt did not emphasise the value of Avogadro's hypothesis. Nor did he, in deriving his table of atomic weights, apply Dulong and Petit's Law. He regarded all metal oxides as of the type M_2O. Consequently, the atomic weights that he assigned to silver and the alkali metals were correct but those to many other metals were only half the true values which moved Berzelius to stigmatise Gerhardt's atomic weights as "false".

References p. 117

Laurent[4] (1808–1853), a close collaborator of Gerhardt's saw more clearly the implication of Avogadro's hypothesis (which he called Ampère's). Like Gaudin, he recognized the divisibility of molecules of the elements such as hydrogen, chlorine and oxygen and wrote equations like:

$$HH + ClCl = HCl + HCl$$

$$HH + HH + OO = HHO + HHO$$

"We admit" he said "that each molecule of a simple body is divisible into at least two parts which we call atoms; these molecules can be divided only in the case of combination". Despite the soundness of many of their ideas, Laurent and Gerhardt aroused the mistrust and opposition of many chemists.

So intolerable was the confusion that prevailed among chemists towards the end of the 1850's that Kekulé and a group of other leading chemists decided to call an international meeting of chemists for the express purpose of obtaining agreement on the following points: precise definitions of the ideas conveyed by the words atom, molecule, equivalent, atomicity, basicity; examination of the true equivalents of substances and their formulae; initiation of a uniform nomenclature and a rational nomenclature. It may seem strange that, as is evident from the agenda, chemists should have taken so long to agree upon definitions of atom, molecule, equivalent and valency. The key to understanding these terms was an unequivocal method for determining the true relative weights of atoms.

On the morning of September 3, 1860, about 140 of the most well-known chemists of the day assembled at Karlsruhe (Germany). In opening the Conference, C. Weltzien said (in part): "For the first time we have here assembled the representatives of a single science and the youngest. The representatives belong to almost all nationalities. We belong to different national groups and speak different languages, but we are united by a scientific interest, by a common intention. We have assembled for the purpose of working towards an agreement on a few important points in the interests of our illustrious science. With the extraordinarily rapid development of chemistry, especially with the enormous accumulation of factual material, the theoretical views of the investigators as shown in their expressions and in the symbols used have diverged more than is expedient for mutual understanding and more than is practical for teaching"[5].

It would take too long to try to follow the discussions. Suffice it to say that to its organisers, the Congress must have seemed a failure. There was a great deal of discussion on the precise meanings to be attached to the terms atom, molecule and equivalent. The main problem, that of devising a rational and unified system of atomic weights, was not solved. Looking back on it, E. von Meyer commented that evidently members came insufficiently prepared for the work of the Congress and that most of the impromptu speeches given during the sessions were to a large extent unhelpful. Dumas, for example, advocated the view that there are two chemistries—inorganic and organic and that equivalents should be used in the former and atomic weights in the latter. The only man who came to the Congress adequately prepared for its business was Cannizzaro (1826–1910) an Italian professor of chemistry who was also a great teacher. He spoke of the work of his fellow countryman, Avogadro, as well as that of Dumas, Gaudin and Gerhardt. "No matter what you do" he said "you will not be able to prevent Gerhardt's system from daily winning new adherents". But he failed to convert his listeners. At the close of the meeting Angelo Pavesi, professor of chemistry at the University of Pavia and a friend of Cannizzaro distributed reprints of the latter's now famous paper[6] "*Sunto di un corso di filosofia chimica*". In this Cannizzaro clearly expounded a rational method for determining the relative weights of atoms based on the hypothesis of Avogadro.

In the next few years, the fundamental ideas expressed in this paper were gradually and finally almost universally accepted. Conversion in many instances was slow—in others, almost dramatically sudden. Thus L. Meyer who read the pamphlet on his return journey to Breslau wrote "It was as though the scales fell from my eyes, doubt vanished and was replaced by a feeling of peaceful certainty".

What was the nature of Cannizzaro's revolutionary paper? It was a description of a university course in chemistry in which he explained how he introduced students to the study of chemistry. He began by sketching the history of chemical theories—the work of Gay-Lussac, Avogadro, Dumas, Laurent, Gerhardt and others. From this survey, he reached the conclusion that "to bring harmony into all the branches of chemistry we must have recourse to the complete application of

References p. 117

Avogadro's hypothesis". In his fifth lecture, he did just this. He not only revived Avogadro's hypothesis but he applied it in an entirely novel way.

In 1811, Avogadro had shown on the assumption that equal volumes of gases under the same conditions of temperature and pressure contain the same number of molecules how it is possible to determine the relative weights of atoms by determining the relative densities of the common elementary gases. By this method, he measured the relative weights of the atoms of hydrogen, oxygen, nitrogen and chlorine, all of which exist in the gaseous state as diatomic molecules.

Early in his fifth lecture Cannizzaro made a simple but important modification to Avogadro's system of calculation. It will be recalled that Avogadro adopted as his standard the density of hydrogen gas $= 1$ which implied that $H = 1/2$.

"Since I prefer" said Cannizzaro "to take as the common unit for the weights of molecules and their fractions, the weight of half and not the whole of a molecule of hydrogen, I therefore refer the densities of the gaseous bodies to that of hydrogen $= 2$" (*i.e.* $H = 1$). To illustrate this point he drew up a table part of which is reproduced below:

TABLE 6

GAS DENSITY STANDARDS

Name of substance	Weights of molecules referred to	
	$H_2 = 1$*	$H = 1$**
Hydrogen	1	2
Oxygen	16	32
Chlorine	35.5	71
Mercury	100	200
Water	9	18

* Density of hydrogen $= 1$.
** Density of hydrogen $= 2$.

Cannizzaro adopted a new approach which was made possible by the greatly increased number of compounds that had been discovered and studied in the 50 years since the first statement of Avogadro's hypothesis. Most numerous among these were the compounds of carbon, many of which could be vaporised without decomposition at temperatures not greatly above room temperatures. This meant that they lent themselves to vapor density measurements by the Dumas method.

Whereas Avogadro in 1811 had measured atomic weights, by determining the relative densities of the common elementary gases, Cannizzaro now examined a whole series of compounds of a particular element whose atomic weight was to be determined. We will consider, as he did, compounds of hydrogen first and use his convention of making the density of hydrogen = 2. See Table 7:

TABLE 7

THE ATOMIC WEIGHT OF HYDROGEN*

The first column of figures shows the weight in grams of a unit volume of the gas concerned. Any unit of volume may be used but for convenience, we will use the volume (22.4 l) occupied by 2.016 g of hydrogen at STP. The next column shows the results of chemical analysis of the gas—stated as the weight in grams of the element concerned that is present in 22.4 l of the gas. For hydrogen compounds, this weight is 1.008 g or some integral multiple of 1.08 g**.

Substance	Wt. (g) of 22.4 l at STP	Wt. (g) of hydrogen in 22.4 l	
Hydrochloric acid	36.5	1.008	1×1.008
Hydrobromic acid	81	1.008	1×1.008
Hydriodic acid	128	1.008	1×1.008
Water	18	2.016	2×1.008
Hydrogen sulphide	34	2.016	2×1.008
Hydrogen	2	2.016	2×1.008
Ammonia	17	3.024	3×1.008
Hydrogen phosphide	34	3.024	3×1.008
Methane	16	4.032	4×1.008
Ethane	30	6.048	6×1.008
		G.C.M.	1.008

* Cannizzaro, like Dalton, made the atomic weight of hydrogen the reference point of the scale of atomic weights by assigning it the value 1. On this scale O = 15.87. In 1884 Marignac suggested that it would be more convenient to refer all atomic weights to that of oxygen giving it the value 16.00. The adoption of this standard was recommended by an international commission in 1900. On the scale of O = 16.00, H = 1.008. Illustrating Cannizzaro's principle it seemed preferable to use the more modern system of atomic weights.

** It is assumed that the molecule of each of the three compounds with the least hydrogen contains one atom. If this is so, then the molecule of elementary hydrogen is diatomic, a fact established in the first instance by Avogadro himself by consideration of the volume relations of the chemical combination of gases.

When a series of oxygen compounds is treated in the same way (Table 8), it will be seen that each contains a mass of oxygen which is 16 or an integral multiple of 16—that the molecule of oxygen con-

TABLE 8
THE ATOMIC WEIGHT OF OYYGEN

Substance	Wt. (g) of 22.4 l at STP	Wt. (g) of oxygen in 22.4 l	
Water	18	16	1×16
Carbon monoxide	28	16	1×16
Phosphorus oxychloride	153.5	16	1×16
Nitric oxide	30	16	1×16
Oxygen	32	32	2×16
Carbon dioxide	44	32	2×16
Sulphur dioxide	64	32	2×16
Chlorine peroxide	67.5	32	2×16
Sulphur trioxide	80	48	3×16
Methyl nitrate	77	48	3×16
Osmium tetroxide	255	64	4×16
		G.C.M.	16

TABLE 9
THE ATOMIC WEIGHT OF CARBON

Compound	Wt. (g) at 22.4 l at STP	Wt. (g) of carbon 22.4 l	
Methane	16	12	1×12
Chloroform	119.5	12	1×12
Carbon monoxide	28	12	1×12
Carbon dioxide	44	12	1×12
Cyanogen chloride	61.5	12	1×12
Ethylene	28	24	2×12
Ethane	30	24	2×12
Cyanogen	52	24	2×12
Acetylene	26	24	2×12
Propane	44	36	3×12
Butane	58	48	4×12
Pentane	72	60	5×12
Hexane	86	72	6×12
Benzene	78	72	6×12
		G.C.M.	12

tains twice this amount and is therefore diatomic. Similar treatment of carbon compounds (Table 9) shows that every compound contains 12 or an integral multiple of 12 g of carbon.

On the basis of data of this kind Cannizzaro[6] made the following generalisation.

"The different quantities of the same element contained in different molecules are all whole multiples of one and the same quantity which, always being entire, has the right to be called an atom".

Stanislao Cannizzaro

The atomic weight of an element is then the least weight of it contained in a molecular weight of a volatile compound. This operational definition of atomic weight which is Cannizzaro's great contribution to the atomic theory assumes that, if a sufficiently large number of compounds of any particular element is considered, one at least will contain only one atom of the element under consideration*.

* The same kind of procedure was used by R. Millikan in measuring the charge of an electron. He found that the charge on an oil drop was always some multiple n of the charge 4.8×10^{-10} e.s.u.

References p. 117

Surprising as it may seem, one of the questions debated at the first meeting of the Karlsruhe Conference[20] was: "Is it convenient to distinguish between atoms and molecules?" Since the meeting adjourned without reaching agreement on what the answer to this question should be, it is reasonable to conclude that the Conference was not able to agree on how, conveniently or otherwise, to distinguish between atoms and molecules.

Cannizzaro's work made the distinction abundantly clear: atoms are the units from which molecules are built, or as Kekule once put it: "atoms are the particles of matter that undergo no further subdivision in chemical transformations". A more formal definition was provided by Roscoe in a textbook[7] he wrote several years after attending the Karlsruhe Conference: "An atom is the smallest portion of an element which can enter a chemical compound". Although Dalton did not use exactly these words, they express what he had in mind.

At the same time as he defined an atom, Roscoe[7] also defined a molecule: it "is the smallest portion of a simple or compound body capable of existing in the free state". He did not elaborate on what he meant by the free state. So far as Cannizzaro's principle is concerned, free state means the gaseous state where molecules are separated from one another by distances which, on the average, are large compared with their own dimensions. A few years later, the concept of free state in this context was extended to cover molecules in solution. Van 't Hoff[8] showed that there is a very close analogy between the behaviour of molecules in a gas and the behaviour of solute molecues in a dilute solution where they are separated from one another by distances which, on the average, are larger than their dimensions. The osmotic pressure of a dissolved substance is the analogue of the pressure of a gas and is proportional to the number of solute molecules, of whatever kind, per unit volume and the absolute temperature of the solution. It thus became possible, by making use of measurement of osmotic pressure and various other colligative properties of solutions, to measure the molecular weights of dissolved substances. This development was a particularly valuable one because it made it possible to measure the molecular weights of the many substances that cannot be volatilised without undergoing decomposition.

When he came to deal with the metallic elements, Cannizzaro could, as a rule, no longer apply Avogadro's hypothesis since not enough stable, volatile metal compounds were known at the time. He fell back on Dulong and Petit's Law. By this time, the specific heats of most of the known metals had been determined and Cannizzaro applied the law systematically as Table 10 taken from his 1858 paper shows.

TABLE 10

ATOMIC HEATS (CANNIZZARO, 1858)

Substance	At. wt.	Spec. heat	At. wt. × spec. heat
Mercury (solid) (Hg)	200	0.0324	6.482
Copper (Cu)	63	0.0951	6
Zinc (Zn)	66	0.0955	6.613
Lead (Pb)	207	0.0314	6.500
Iron (Fe)	56	0.1138	6.372
Manganese (Mn)	55	0.1181	6.495
Tin (Sn)	117.6	0.0562	6.612
Platinum (Pt)	197	0.0324	6.388
Potassium (K)	39	0.1695	6.613
Sodium (Na)	23	0.2934	6.748
Silver (Ag)	108	0.0570	6.157

At this period there were not many opportunities for making a direct cross-check on the two methods for determining atomic weights for the simple reason that the number of elements to which Cannizzaro had applied his method was not great*. It is probable that vapor density measurements had been made on enough sulphur compounds for sulphur to be used for this purpose but Cannizzaro did not attempt to do this.

Cannizzaro also made use of the relationship discovered by Kopp[10]: the molecular heat of a solid compound $A_x B_y C_z$ is the sum of the atomic heats of the elements; if C = spec. heat, a = at. wt. of A then molecular heat is $\sum Cax$ (See Table 11.)

Though the fog around the problem of determining atomic weights had dispersed and it was clear enough to see many things, not every chemist saw clearly. The acceptance of Cannizzaro's ideas was by no means universal or rapid. Nine years after the Karlsruhe Conference,

---

* Some years later (1875) when Kundt and Warburg[9] showed, from velocity of sound measurements, that mercury vapor is monatomic a cross check was possible by using vapor density measurements of the element.

References p. 117

the President of the Chemical Society of London, Williamson[11], felt obliged to give a lecture in defence of the atomic theory. The occasion was his Presidential Address to the Society on June 3, 1869.

TABLE 11

MOLECULAR HEATS (CANNIZZARO, 1858)

Compound	Mol. wt.	Spec. heat	Mol. heat	Number of atoms	Mol. heat/n
KCl	74.5	0.1729	12.884	2	6.442
AgCl	143.5	0.0911	13.071	2	6.535
KBr	119	0.1132	13.473	2	6.736
NaI	150	0.0868	13.026	2	6.513
$HgCl_2$	271	0.0689	18.669	3	6.223
$ZnCl_2$	134	0.1362	18.6567	3	6.2189
HgI_2	454	0.04197	19.6708	3	6.5569

It was a long rambling affair in which he mentioned Cannizzaro once and then only in connection with the application of Dulong and Petit's Law as a means of deciding the atomic weights of metals. He referred at great length to the Law of Multiple Proportions but made no mention of Cannizzaro's principle which provided just as strong evidence in favor of the atomic theory as did that Law. He spoke at considerable length about equivalent weights but made no mention of the new status of atomic weights. In view of the fact that he was one of the signatories to the invitation issued to chemists to attend the Karlsruhe conference this is surprising. Nevertheless he was an unwavering believer in the existence of atoms. His address provoked a sharp reaction among his hearers and became the center of what Knight[12] has referred to as the *Atomic Debates*.

Some of the discussions centered around the question whether atoms were indivisible bodies. Two years before Williamson's address Kekulé had declared:

"I have no hesitation in saying that, from a philosophical point of view, I do not believe in the actual existence of atoms taking the word in its literal signification of indivisible particles of matter. As a chemist, however, I regard the assumption of atoms, not only as advisable but absolutely necessary in chemistry. I will even go further and declare my belief that *chemical atoms exist* provided the term be understood to denote those particles of matter which undergo no further division in chemical metamorphoses".

The reactions to Williamson's address however, went beyond this. There were chemists who denied a belief in the existence of chemical atoms as defined by Kekule. Foremost among them was Brodie, Professor of Chemistry at the University of Oxford. In an address in 1867 to the Chemical Society entitled *"On the Mode of Representation afforded by the Chemical Calculus as contrasted with the Atomic Theory"* Brodie[13] ridiculed the atomic theory and is said to have caused much laughter among his audience by his comments on molecular models. He attributed the use of these models among others to Frankland.

In the ensuing discussion Frankland[14] hotly denied that he used graphical formulae and models to represent the position of atoms in relation to one another in a molecule. He went even further: "perhaps I cannot do better" he said "than state simply and at once that I do not believe in atoms themselves", a statement which is said to have shocked Odling, a well known chemist of that time, who like Williamson had attended the Karlsruhe Conference.

Three years previously Odling had addressed the British Association for the Advancement of Science on the subject of the atomic theory. Though he said he fully admitted the new system of atomic weights, he does not appear to have grasped its basis if one may judge by his comments in that address. Thus Odling[15] declared "equal volumes of all gases, elementary or compound, contain the same number of atoms". The only reference he made to Cannizzaro was to his recognition of diatomic [divalent] metals.

In a paper that appeared in 1871 Mills attacked the atomic theory in no uncertain manner. An interesting argument he used against the theory was one first put forward by Faraday who was also sceptical of the atomic theory. To quote Mills' own words[16]:

"Now of the two constituents of matter (atoms and empty space), space is the only continuous one. Consider then, the case of shellac, a non-conductor. Space in it must be an insulator whatever the atoms may be; for if it were a conductor, the shellac would not insulate. But now take the case of platinum which must also be composed of atoms and space. Since platinum is a conductor, space being its only continuous constituent, must be a conductor. Space, which is everywhere uniform, is therefore a conductor and a non conductor". To-

References p. 117

wards the end of his paper Mills concluded: "And on enquiry it (a logical mind) will find, if my argument be sound, that the atomic theory has no experimental basis, is untrue to nature generally and consists in the main of a materialistic fallacy, derived from appetite more than judgment".

It will be clearly evident from this that the progress of the atomic theory was not a succession of uninterrupted and readily accepted victories. However, despite these and other rearguard actions, Cannizzaro's work gradually became more and more widely accepted.

A factor that contributed to the acceptance of Cannizzaro's self consistent scheme of atomic weights was a slowly growing awareness by chemists of the work of the physicists. Especially important was the evidence provided by the dynamic or kinetic theory of gases. On the basis of the fundamental postulates of this theory and by applying statistical methods to the assembly of molecules constituting a gas Clausius[17] and Maxwell were able to demonstrate the correctness of Avogadro's Hypothesis. This proof, provided by two of the outstanding mathematical physicists of the 19th century greatly strengthened the foundation of Cannizzaro's system of atomic weights. Curiously enough Maxwell[18] when referring to this proof made no mention of Avogadro. He did mention Gay-Lussac and seems to have attributed the hypothesis to him.

The establishment of a rational system of atomic weights had a profound influence on the development of chemistry in two major directions:

(a) on the theory of molecular structure; (b) the discovery of the periodic law.

Before a theory of molecular structure could be developed, it was essential to know the correct formulae of the substances studied and these formulae could only be deduced from a knowledge of the true relative weights of atoms. Without a knowledge of the correct formulae of molecules little progress could be made in understanding the idea of chemical valence.

In an address given in 1956 to commemorate the 100th anniversary of Avogadro's death, Pauling[19] said "We now can see in retrospect, that almost the whole of the development of the science of chemistry has followed from the acceptance of the correct atomic weights and the subsequent development of chemical structure theory".

It was also on the firm foundation of a system of rational atomic weights that Newlands, Meyer and Mendeleeff were able to discover the Periodic Law—another cornerstone of the edifice of chemistry.

REFERENCES

1 A. D. RITCHIE, The Atomic Theory, *Mem. Proc. Manchester Lit. Phil. Soc.*, 86 (1944) 180.
2 A. KEKULE, *Lehrbuch der organischen Chemie*, Erlangen und Stuttgart, 1866.
3 C. F. GERHARDT, *Ann. Chim.*, 8 (1843) 129, 238.
4 A. LAURENT, *Ann. Chim.*, 18 (1846) 266.
5 C. DE MILT, Carl Weltzien and the Congress at Karlsruhe, *Chymia*, 1 (1948) 153.
6 S. CANNIZZARO, *A Sketch of a Course of Chemical Philosophy*, Alembic Club Reprint No. 18, Livingstone Ltd., Edinburgh, 1947.
7 H. E. ROSCOE, *Lessons in Elementary Chemistry*, Macmillan and Co., London. First edition 1866, Seventh edition 1900. The reference is to the latter, p. 59.
8 J. H. VAN 'T HOFF, *Z. Phys. Chem.*, 1 (1887) 481.
9 A. KUNDT AND E. WARBURG, *Ber.*, 8 (1875) 945.
10 H. KOPP, *Ann. Chim.*, 126 (1863) 362.
*11 A. WILLIAMSON, On the Atomic Theory, *J. Chem. Soc.*, 22 (1869) 328.
12 D. KNIGHT, *The Atomic Debates*, Leicester University Press, 1967.
*13 B. C. BRODIE, *Chem. News*, 15 (1867) 295.
*14 E. FRANKLAND, *Chem. News*, 20 (1869) 235.
*15 W. ODLING, *Report to the British Association*, 1864, p. 21.
*16 E. J. MILLS, *Phil. Mag.*, 42 (1871) 112.
17 R. J. E. CLAUSIUS, *Ann. Phys.*, 100 (1857) 353.
*18 J. W. MAXWELL, *Encyclopaedia Britannica*, Vol. 3 (1875), p. 36.
19 L. PAULING, *Science*, 124 (1956) 710.
20 H. HARTLEY, *Notes and Records of the Royal Society of London*, 21 (1966) 56.

* The papers marked with an asterisk have been reprinted in D. KNIGHT (Ed.), *Classical Scientific Papers*, Mills and Boon, London and Elsevier Publishing Co., Amsterdam, 1968.

CHAPTER 8

Consequences of solving the problem of atomic weights (I). Discovery of the periodic law

"No law of nature, however general, has been established all at once; its recognition has always been preceded by many presentiments. When one considers carefully the genesis of any discovery", wrote Sarton, the Harvard historian of science, "one finds that it was always prepared by a number of smaller ones, and the deeper one's investigations, the more intermediary stages are found". This is certainly true of the discovery of the Periodic Law, one of the most comprehensive and useful of all generalisations in chemistry.

The discovery of the Periodic Law, which states that the chemical and physical properties of the elements are a periodic function of their atomic weights (or as we should now say, atomic numbers), depended on three antecedents:

(1) the development of the concept of an element;
(2) the recognition of the existence of families of similar elements;
(3) the establishment of the true relative weights of atoms of the elements.

Classification of substances has long been a preoccupation of chemists. Lavoisier took an important step in this direction when, in 1789, he classified the 33 substances he had listed as elements. Light and heat (then thought to be material) together with the 3 gases oxygen, hydrogen and nitrogen formed the first class; the 17 metals formed the second; six oxidizable non-metals the third and the fourth comprised the earths which were subsequently shown to be oxides. The number of elements known to him was so small that there was not much opportunity for him to carry the classification further. However, in the next 80 years, the number of elements brought to light was considerable.

In the course of discovering these elements and in the investigation of their chemical and physical properties and those of their compounds, close resemblances between varous elements were revealed, causing them to be grouped in families. By about 1860, the 57 known elements had been arranged in families of varying size up to 6. Table 12 shows a grouping compiled by Odling[1] in 1861. Some of the elements shown would not now be so grouped but the halogen, oxygen, nitrogen, carbon (except for tantalum), alkali metal, alkaline earth and platinum metal families are as accepted today.

TABLE 12

FAMILIES OF ELEMENTS (ODLING, 1861)

H	Hydrogen	1	Mg	Magnesium	12
			Zn	Zinc	32.5
F	Fluorine	19	Cd	Cadmium	56
Cl	Chlorine	35.5			
Br	Bromine	80	Hg	Mercury	100
I	Iodine	127	Pb	Lead	103.5
			Ag	Silver	108
O	Oxygen	16			
S	Sulphur	32	Cr	Chromium	26
Se	Selenium	80	Mn	Manganese	27
T	Tellurium	128	Fe	Iron	28
			Ni	Nickel	29
N	Nitrogen	14	Co	Cobalt	30
P	Phosphorus	31	Cu	Copper	31.7
As	Arsenic	75			
Sb	Antimony	120			
Bi	Bismuth	208	Al	Aluminum	13.7
			Zr	Zirconium	33.5
C	Carbon	12	Ce	Cerium	46
Si	Silicon	28.5	La	Lanthanum	47
Ti	Titanium	48.5	D	Didymium	48
Sn	Tin	118	U	Uranium	60
Ta	Tantalum	138			
			Mo	Molybdenum	48.
B	Boron	11	Vd	Vanadium	68.5
			W	Tungsten	92
Li	Lithium	7			
Na	Sodium	23	Au	Gold	197
K	Potassium	39			
			Ro	Rhodium	52
Ca	Calcium	20	Ru	Ruthenium	52
Sr	Strontium	44	Pd	Palladium	53
Ba	Barium	68.5			
			Pr	Platinum	98.5
G	Glucinum	4.7	Ir	Iridium	98.5
Y	Yttrium	32	Os	Osmium	99.5
Th	Thorium	59.5			

The first supposed relationship between the atomic weights of the elements was that suggested in 1815 by a medical practitioner, Prout[2] (1785–1850), who believed that the atomic weights of the elements were integral multiples of that of hydrogen (taken as unity). Thomson (mentioned earlier as the first to publish an account of Dalton's atomic theory) took up Prout's idea. In his book Thomson[3] produced a table of atomic weights, some of which he had himself determined, in support of Prout. It happened that he had corrected several of the values of Berzelius. This so angered the latter that he accused Thomson of what amounted to fraud. In a review of Thomson's book, Berzelius wrote "it belongs to those few productions from which science will derive no benefit whatever. Much of the experimental part, even of the fundamental experiments, appears to have been made at the writing desk; and the greatest civility which his contemporaries can show its author, is to forget it was ever published". All the evidence points to the fact that Thomson was an honest man. It seems that a method he used for estimating the atomic weight of zinc, one of the elements whose atomic weight was in dispute, was one that has since been found to give inaccurate results.

Partington[4] has drawn up an interesting table (Table 13) comparing some of the atomic weights of Thomson (I) and Berzelius (II) compared with the modern values (III) (the two former being recalculated to the value $O = 16$).

TABLE 13

ATOMIC WEIGHTS AND PROUT'S HYPOTHESIS

	H	Cl	C	N	S	Ag	Hg	Ca
I	1	36	12	14	32	108	200	40
II	0.998	35.41	12.23	14.16	32.19	108.12	202.5	40.96
III	1.008	35.46	12.01	14.008	32.07	107.88	200.61	40.08

One result of Prout's hypothesis was to stimulate chemists to make more and more accurate determinations of atomic weights with the result that the hypothesis was discredited though it was undeniable that more atomic weights closely approximated whole numbers than could be attributed to chance*.

One of the earliest attempts to correlate the properties of elements with their atomic weights was that of Döbereiner[5] who in 1817 pointed out that the atomic weight of strontium (42.5) is the arithmetic mean of those of calcium (20) and barium (65)**. It will be seen of course that the numbers he used are equivalent weights which in each case is half the atomic weight. This, however, did not obscure a relationship. About 10 years later he discovered a similar relationship between what were really the atomic weights of other groups of three elements which became known as triads.

He says "The work of Berzelius on the determination of the atomic weights of bromine and iodine interested me greatly, since it has established the idea, which I expressed earlier in my lectures, that perhaps the atomic weight of bromine might be the arithmetical mean of the atomic weights of chlorine and iodine. This mean is

$$\frac{35.470+126.470}{2} = 80.470 \text{ (sic)}$$

This number is not much greater [his arithmetic here was not very good, or a printer's error may be involved] than that found by Berzelius (78.383); however, it comes so close that it may almost be hoped that the difference will vanish entirely after repeated careful and exact determination of the atomic weights of these three salt forming elements".

The values he used for sulphur and tellurium in his paper were

$$\frac{32.239 \,(= S) + 129.234 \,(= Te)}{2} = 80.736$$

The value then in use for selenium was 79.263. Modern determinations of atomic weights show that chemists of that period were far more confident of the accuracy of their measurements than was justified. Nevertheless Döbereiner clearly demonstrated a relationship between

* While it is true that Prout's ideas about atomic weights were not borne out by experiment, modern work has shown that his notion that all elements were built up from hydrogen had a germ of truth in it. It is now believed that the elements have indeed been and are being synthesised from hydrogen. The sun derives its energy from thermonuclear reactions in which hydrogen is converted to helium.

** He first noted that the molecular weights of the oxides of the three elements were related in this way; also those of the sulphates.

References pp. 133-134

atomic weights of chemically similar elements. His discovery attracted much attention and set chemists off on a search for other arithmetic relationships between the atomic weights of similar elements.

An interesting relationship and one that contained a number that was later to prove of great significance was turned up in 1850 by Pettenkofer[6] who noticed that the atomic weights of similar elements often differed from one another by a multiple of eight

$$\begin{array}{ll} \text{Li} = 7 & \text{Mg} = 12 \\ \text{Na} = 7+16 & \text{Ca} = 12+8 = 20* \\ \text{K} = 23+16 & \text{Sr} = 24+20 = 44* \\ & \text{Ba} = 44+24 = 68* \end{array}$$

Without mentioning the work of either Döbereiner and Pettenkofer, Dumas in a lecture to the British Association for Advancement of Science at Ipswich (1851), pointed out the following relationship:

$$\begin{array}{l} \text{N} = 14 \\ \text{P} = 14+17 = 31 \\ \text{As} = 14+17+44 = 75 \\ \text{Sb} = 14+17+88 = 119 \\ \text{Bi} = 14+17+176 = 207 \end{array}$$

Up to this time chemists had been looking at the atomic weights of small groups of chemically similar elements when placed in increasing order. It occurred to Gladstone to consider the atomic weights of all the known elements (56) when placed in order of their increasing values. Unfortunately for him, any opportunity he might have had of discovering any relationship was frustrated by the fact that he used atomic weights for some elements and equivalent weights for others (mainly the latter).

It is significant that this approach—looking at all the known elements arranged in order of increasing atomic weights—began to produce results soon after the acceptance of the results of Cannizzaro's work.

Then followed what is by no means a rare occurrence in the history of science. The law, known today as the Periodic Law, was discovered independently, though not simultaneously, by six individuals: Béguyer

---

* It should be noted that the then accepted values of the atomic weights of these elements were still half the modern one. However, as will already have been noted, Berzelius' values for Cl Br I, S Se Te, approximate to modern values.

de Chancourtois (France), Newlands and Odling (England), Hinrichs (U.S.A.), Meyer (Germany) and Mendeleeff (Russia)[7].

In 1862, 2 years after the Karlsruhe Conference, a French mineralogist, Béguyer de Chancourtois[18] (1819–1866), arranged the elements in order of increasing atomic weight on a spiral, generated on the surface of a cylinder, in such a way that similar elements fell on the same vertical line. To the spiral he gave the name *Vis Tellurique* ("telluric helix"). As a result of arranging the elements in this way he reached the conclusion that "the properties of bodies (elements) are the properties of numbers". For some reason that is only partly clear, this work remained practically unknown to chemists until about 1889. In the first publication describing his work the editors of the journal (*Comptes Rendus*) omitted the accompanying illustration which must have increased the difficulty of understanding his ideas. He later published an account of his system privately in a pamphlet but this was not widely circulated.

The second discovery and probably the most controversial—was made by Newlands (1837–1898), a young English chemist then aged only 25, who had recently returned from fighting in Italy under Garibaldi. After at least three attempts to discover significant relationships between atomic weights and properties of the elements, Newlands[8] published in a journal that no longer appears (*Chem. News*) a paper which contained the germ of the Periodic Law. His first attempts were frustrated partly at least because of the confusion between equivalent and atomic weights. He said that he "had tried several other schemes before arriving at the one now proposed. One founded upon the specific gravity of the elements failed altogether and no relation could be worked out of the atomic weights under any other system than that of Cannizzaro". "If", said Newlands, "the elements are arranged in order of their equivalents [atomic weights] with a few slight transpositions, as in the accompanying table (Table 14) it will be observed that elements belonging to the same group usually appear on the same horizontal line. It will also be seen that the number of analogous elements generally differ either by 7 or some multiple of 7*; in other words members of each group stand to each other in the same relation

---

* The noble gases of the atmosphere—helium, neon, *etc.*—had not then been discovered.

as the extremities of one or *more octaves* in music. Thus, in the nitrogen group between nitrogen and phosphorus there are seven elements; between phosphorus and arsenic 14; between arsenic and antimony 14 and lastly between antimony and bismuth 14 also. This peculiar relationship I propose to term provisionally "The Law of Octaves". (See also ref. 9.)

TABLE 14

THE LAW OF OCTAVES (NEWLANDS, AUGUST 1865)

El.	No.	El.	No.	El.	No.	El.	No.	El.	No.	El.	No.	El.	No.
H	1	F	8	Cl	15	Co, Ni	22	Br	29	Pd	36	I	42
Li	2	Na	9	K	16	Cu	23	Rb	30	Ag	37	Cs	44
G	3	Mg	10	Ca	17	Zn	25	Sr	31	Cd	38	Ba, V	45
Bo	4	Al	11	Cr	19	Y	24	Ce, La	33	U	40	Ta	46
C	5	Si	12	Ti	18	In	26	Zr	32	Sn	39	W	47
N	6	P	13	Mn	20	As	27	Di, Mo	34	Sb	41	Nb	48
O	7	S	14	Fe	21	Se	28	Ro, Ru	35	Te	43	Au	49

Two years later he modified this table slightly by improving the positions of the thallium, lead and mercury.

On March 1, 1866 he presented a paper containing the improved version of Table 14 to the Chemical Society (London). It was received with a great deal of scepticism—even ridicule, and though submitted for publication was never published. So little did the chemists present on that occasion appreciate Newlands' work that one critic, Professor G. C. Foster, is reported to have humorously asked Newlands whether he had ever thought of arranging the elements according to the initial letters of their names since any arrangement would present occasional coincidences.

About seven years later when Newlands sent a note to the *Journal of the Chemical Society* with the object of establishing the priority of his claim as discoverer of the Periodic Law he was again refused publication and told by the President of the Society, Dr. W. Odling the reason for the refusal of the first paper. He stated that the Society "had made it a rule not to publish papers of a purely theoretical nature, since it was likely to lead to correspondence of a controversial character". Newlands was never elected a Fellow of the Royal Society though he was awarded the Society's Davy Medal in 1887, a delayed recognition which probably more than made amends for his being passed over for election to a Fellowship.

Newlands undoubtedly established several fundamental ideas: (1) the concept of atomic number (he did not use this term exactly—he spoke of the number of the element) this being the ordinal number when the elements are arranged in increasing order of atomic weights. (2) the repetition of similar properties after a regular interval when the elements are so arranged, in other words that the chemical and physical properties of the elements are a periodic function of their atomic weights. At one stage he did suggest that there were missing elements but he did not follow this up.

Before turning to the work done on the Continent, mention must be made of a remarkable table compiled by Odling who, it will be recalled, had a very good knowledge of the way in which elements could be arranged in groups.

In October 1864 Odling[10] published a paper entitled "*Proportional Numbers of the Elements*" which one historian[11] says was "probably prepared without a knowledge of Newlands' work". In this he listed 61 elements in order of increasing atomic weight and showed that "this purely arithmetic seriation may be made to accord with a horizontal arrangement of the elements according to their usual received groupings" (see Table 15).

TABLE 15

AN ARRANGEMENT OF THE ELEMENTS (ODLING, OCTOBER 1864)

					Ro 104	Pt 197
					Ru 104	Ir 197
					Pt 106.5	Os 199
H 1	,,	,,	Zn 65	Ag 108		Au 196.5
,,	,,	,,		Cd 112		Hg 200
L 7	,,	,,		,,		Tl 203
G 9	,,	,,		,,		Pb 207
B 11	Al 27.5	,,		U 120		,,
C 12	Si 28	,,		Sn 118		,,
N 14	P 31	,,	As 75	Sb 122		Bi 210
O 16	S 32	,,	Se 79.5	Te 129		,,
F 19	Cl 35.5	,,	Br 80	I 127		,,
Na 23	K 39	,,	Rb 85	Cs 133		,,
Mg 24	Ga 40	,,	Sr 87.5	Ba 137		,,
	Ti 50	,,	Zr 89.5	Ta 138		Th 231.5
	,,		Ce 92	,,		
	Cr 52.5		Mo 96	V 137		
	Mn 55			W 184		
	Fe 56					
	Co 59					
	Ni 59					
	Cu 63.5					

References pp. 133-134

It will be seen in the sequel that this table bears a remarkable resemblance to one published a few years later in Russia.

Within three or four years of Newlands' statement of the Law of Octaves, two other chemists, namely, the German, Lothar Meyer, (1830–1895) and the Russian, Dimitri Mendeleeff, announced they had discovered a periodic relation between the atomic weights and chemical properties of the atoms. It is interesting to note that each

Dimitri Ivanovitch Mendeleeff

developed the discovery in the course of writing a textbook: Meyer—*Modern Theories of Chemistry*, and Mendeleeff—*Foundations of Chemistry*. Mendeleeff was the first of the two to publish a periodic table[13–15]*. Meyer[12] did not publish his table until 1870, though he is said to have made some attempt at classification in the first (1864) edition of his text book. It is generally believed that the two men arrived at their discoveries independently.

It is true that a Russian biographer[16] of Mendeleeff hints that Meyer knew about Mendeleeff's work before it appeared in print. However, Partington cites documentary evidence to show that Meyer drew up a periodic table in July 1868 which was intended for a new edition of his book[12]. The date of publication is of course the safe way of establishing priority but independent discovery is not necessarily ruled out by later publication. Mendeleeff has said that at the time he first announced his discovery of the Periodic Law he was unaware of the work of Newlands and also that of Meyer.

Mendeleeff's name is usually most prominently associated with the Periodic Law. This can be justified without any derogation of the work of his contemporaries or earlier investigators. Here, in his own words[17], is a brief account of the manner in which he made his discovery.

"The decisive moment in the development of my theory of the periodic law was in 1860, at the conference of chemists in Karlsruhe, in which I took part, and at which I heard of the ideas of the Italian Chemist, Cannizzaro. I regard him as my immediate predecessor, because it was the atomic weights which he found, which gave me the necessary reference material for my work. I noted immediately that the modifications he proposed to the atomic weights introduced new patterns into Dumas' groupings and it was then that I was struck with the essential idea of a possible periodicity in the properties of the elements on increase in the atomic weight. I was still hindered by the incongruities in the atomic weights accepted at this time; but I was firmly convinced that this was the direction in which to pursue my work".

"In undertaking to write a text book called *Principles of Chemistry*,"

---

* On 18th March, 1869, at a meeting of the Russian Chemical Society, Professor N. A. Menshutkin read for Mendeleeff who was absent through indisposition his paper entitled *"An Outline of a System of the Elements Based on their Atomic Weights and Chemical Affinities"*.

Mendeleeff wrote, "I wished to establish some sort of system of simple bodies [elements] in which their distribution is not guided by chance, as might be thought instinctively, but by some sort of exact principle. Any system that is based on accurately observed numbers is, of course, for that reason alone preferable to other systems that have no numerical basis, since there will be less room in it for arbitrariness.... The numerical data relevant to simple bodies are at present limited.....

Optical properties, for instance, and even electrical and magnetic properties cannot, of course, serve as the basis of a system because one and the same body can show great differences in this respect according to the state in which it happens to be. It suffices to recall in this respect, graphite and diamond, ordinary and red phosphorus, oxygen and ozone. Yet each of us realises that with all the changes in the properties of simple bodies in a free state, something remains constant... In this connection, we know only one numerical datum—the atomic weight peculiar to the element. This is why I have tried to found a system of classification according to the atomic weights of the elements". (See ref. 16, p. 73.)

His first published table[13] which was circulated to Russian physicists and chemists in February, 1869, is shown in Table 16.

Mendeleeff later was quite ready to admit that the idea of a general law had been foreshadowed by others.

"I see now clearly that de Chancourtois, Newlands and others stood foremost in the way towards the discovery of the Periodic Law and that they merely wanted the boldness necessary to place the whole question at such a height that its reflection on the facts could be clearly seen".

It is the remarkable ability which Mendeleeff showed in the development and application of his theory which distinguishes him from all other workers in this field and entitles him to the highest recognition.

Two developments will be singled out for special mention: (1) the confidence with which Mendeleeff predicted the existence of elements not then discovered; (2) the firmness with which he rejected atomic weights which did not place the elements in what he believed to be their right pigeonholes.

To take the latter point first—the atomic weight of indium was then believed to be 74 but Mendeleeff thought it should be nearer to 114. Further work on the atomic weight of indium proved that Mendeleeff was right. Similarly he suggested that the then listed atomic weights of beryllium, uranium and gold were incorrect and further work resulted in corrections which accorded with the position of these elements in Mendeleeff's table. He was wrong in his views about the order of the

TABLE 16

MENDELEEFF'S "OUTLINE OF A SYSTEM OF THE ELEMENTS"

ОПЫТЪ СИСТЕМЫ ЭЛЕМЕНТОВЪ,

ОСНОВАННОЙ НА ИХЪ АТОМНОМЪ ВѢСѢ И ХИМИЧЕСКОМЪ СХОДСТВѢ.

			Ti = 50	Zr = 90	? = 180.
			V = 51	Nb = 94	Ta = 182.
			Cr = 52	Mo = 96	W = 186.
			Mn = 55	Rh = 104,4	Pt = 197,4.
			Fe = 56	Ru = 104,4	Ir = 198
			Ni = Co = 59	Pl = 106,6	Os = 199.
H = 1			Cu = 63,4	Ag = 108	Hg = 200
	Be = 9,4	Mg = 24	Zn = 65,2	Cd = 112	
	B = 11	Al = 27,4	? = 68	Ur = 116	Au = 197?
	C = 12	Si = 28	? = 70	Sn = 118	
	N = 14	P = 31	As = 75	Sb = 122	Bi = 210?
	O = 16	S = 32	Se = 79,4	Te = 128?	
	F = 19	Cl = 35,5	Br = 80	I = 127	
Li = 7	Na = 23	K = 39	Rb = 85,4	Cs = 133	Tl = 204
		Ca = 40	Sr = 87,6	Ba = 137	Pb = 207.
		? = 45	Ce = 92		
		?Er = 56	La = 94		
		?Yt = 60	Di = 95		
		?In = 75,6	Th = 118?		

atomic weights of tellurium and iodine. All subsequent investigations of the atomic weights of these two elements agree in making the atomic weight of tellurium greater than that of iodine which places these elements in the wrong pigeonholes—wrong, that is, if atomic weights are used as a basis for arranging the elements. We now know, of course, that the charge on the nucleus of an atom or its atomic number is the significant factor and that when this is employed as a basis for arranging

References pp. 133–134

the elements, tellurium and iodine fall into their appropriate places*.

It was Mendeleeff's prediction of the existence of undiscovered elements that did most to make chemists sit up and take notice of his work. He predicted the existence of three elements which he called eka-aluminium, eka-boron and eka-silicon. The positions of these elements in the Periodic Table are indicated by question marks in Table 16. Not only did he predict the existence of these elements but he also predicted with considerable accuracy a number of their chemical and physical properties**.

One example of his predictions must suffice.

TABLE 17

PHYSICAL AND CHEMICAL PROPERTIES PREDICTED FOR GERMANIUM

Mendeleeff's (1871) forecast for eka-silicon	Winkler's (1885) experimental findings on germanium
At. wt. 72	At. wt. 72.6
Spec. gravity of the element 5.5	Spec. gravity 5.35
Will not displace hydrogen from acids	Does not dissolve in acids
The formula for its oxide EsO_2	The formula for its oxide is GeO_2
Sp. gr. of oxide 4.7	The sp. gr. of oxide is 4.7
The chloride $EsCl_4$ will be a liquid b.p. 90°C and sp. gr. 1.9	$GeCl_4$ is a liquid b.p. 83°C and sp. gr. 1.887

Some time after he had discovered germanium Winkler[19] wrote: "I doubt whether there could be a clearer proof of the correctness of the theory of the periodicity of the elements. It is, of course, more than a simple affirmation of a daring theory; it signifies an outstanding broadening of the chemical horizon, a gigantic step forward in the field of knowledge".

* A similar anomaly appeared years later when it was discovered that the atomic weight of argon (39.94) was greater than that of potassium (39.1). From their places in the Periodic Table, one would expect the atomic weight of potassium to be greater than that of argon. The explanation of this anomaly only became clear after the discovery of stable isotopes present in different elements and after their determination by means of the mass spectrometer. Of the three isotopes of K (^{39}K, ^{40}K, ^{41}K) the lightest, ^{39}K, is present to the extent of 93.26 %. On the other hand, with the isotopes of argon (^{36}Ar, ^{38}Ar and ^{40}Ar), the heaviest (^{40}Ar) is present to the extent of 99.57 %.

** Newlands and Meyer also noted the existence of gaps in the Periodic Table but they did not go so far as to predict the physical and chemical properties of the missing elements.

Ever since then the Periodic Law has continued to provide a powerful stimulus and valuable guide in the search for new elements. It would take too long to list them all. Suffice it to say that by about 1939 all the thirty or more gaps in the Periodic Table up to uranium (number 92) had been filled but the search did not end there. The discovery of a very large group of unstable elements known as the trans-uranium elements or actinides owes much to the influence of the Periodic Law.

Mendeleeff himself was, of course, aware of the importance of his own contributions. In 1907, the year of his death, he wrote: "Neither de Chancourtois to whom the French give priority in the discovery of the Periodic Law nor the British candidate, Newlands, nor Meyer, whom some have described as the founder of the Periodic Law, took the risk of forecasting the properties of undiscovered elements, or of changing the accepted weights of atoms and in general did not risk considering the Periodic Law as a new firmly established law of nature capable of embracing facts that till now have not been generalised as I have done in my own work since the beginning (1869)". (See ref. 16, p. 91.)

Mendeleeff's discovery of the Periodic Law was widely acclaimed by many scientists of the time both in his own and other countries. But for all that his experience with what might be called 'official science' was, in some ways, reminiscent of that suffered by Newlands.

On November 11, 1880 the Russian Imperial Academy of Science, the leading scientific organization of the country, held a ballot to decide whether to admit Mendeleeff as a member. By a narrow margin he was rejected. There is some evidence that political rather than scientific considerations influenced the vote. However that may be, the Academy does not seem, in later years, to have recognized his work by any other form of award as did the Royal Society in the case of Newlands.

In later versions of the Periodic Table, Table 18, Mendeleeff turned it around 90° making his vertical columns horizontal and his horizontal columns vertical. The table has undergone many modifications in the arrangement of the elements but the modern form shown in Table 19 is widely used.

It was not long before chemists began to speculate about the implications of the Periodic Law for the atomic theory. They began to wonder whether atoms possessed a structure and whether atoms of elements

References pp. 133-134

TABLE 18

A LATER ARRANGEMENT OF THE ELEMENTS BY MENDELEEFF (1871)

REIHEN	Gruppe I R^2O	Gruppe II RO	Gruppe III R^2O^3	Gruppe IV RH^4 RO^2	Gruppe V RH^3 R^2O^5	Gruppe VI RH^2 RO^3	Gruppe VII RH R^2O^7	Gruppe VIII RO^4
1	H=1							
2	Li =7	Be=9.4	B =11	C =12	N =14	O =16	F =19	
3	Na=23	Mg=24	Al=27.3	Si= 28	P= 31	S= 32	Cl=35.5	
4	K =39	Ca=40	- =44	Ti=48	V =51	Cr=52	Mn=55	Fe=56, Co=59 Ni=59, Cu=63
5	(Cu=63)	Zn= 65	- = 68	- = 72	As= 75	Se=78	Br=80	
6	Rb =85	Sr =87	?Yt=88	Zr =90	Nb=94	Mo=96	- =100	Ru=104, Rh=104 Pd=106, Ag=108
7	(Ag=108)	Cd = 112	In = 113	Sn = 118	Sb= 122	Te= 125	J = 127	
8	Cs =133	Ba=137	?Di=138	?Ce=140	-	-	-	- - - -
9	(-)							
10	-	-	?Er=178	?La=180	Ta =182	W =184	-	Os=195, Ir =197 Pt= 198, Au=199
11	(Au = 199)	Hg=200	Ti= 204	Pb=207	Bi= 208	-	-	
12	-	-	-	Th=231	-	U =240	-	- - - -

TABLE 19

A MODERN VERSION OF THE PERIODIC TABLE

After Von Antropoff, revised by Scheele, 1949. For completeness sake, elements lacking in Scheele's version (97ff) have been added.

showing similar chemical and physical behaviour might not exhibit similar features in their structure. Is it possible, they asked, that atoms are constructed of one or more fundamental building blocks similarities in the arrangement of which could account for similarities in their chemical and physical behaviour?

Mendeleeff, at first, did not accept this view. He wrote[20], "The more I have thought on the nature of the chemical elements the more decidedly I have turned away from the classical notion of a primary matter (protyle) and from the hope of attaining the desired end by a study of electrical and optical phenomena".

He later changed his view. "When, in the future, the time will come to answer such questions as what atomic weight expresses, what is the most likely reason for the dependence of properties on weight, why small changes in the weight of an atom lead to a certain periodic change in properties, then we can expect a theoretical definition of the simplest bodies ... We may easily suppose ... that the atoms of simple bodies are complicated substances, formed by the combination of still more minute particles (ultimates) and what we call indivisible (the atom) is only indivisible by chemical forces—in the way that particles are indivisible by ordinary physical conditions—a theory which though precarious and arbitrary, the chemists' minds involuntarily accept".

The answers given to these questions during the next hundred years were to show that the great contribution to science made by the discovery of the Periodic Law was that it laid the foundation for the view that "the chemical elements are not fragmentary incidental facts of nature but successive units in the sublime harmony of the universe".

REFERENCES

1 W. ODLING, *A Manual of Chemistry*, Part I, Longmans, London, 1861, p. 3.
2 W. PROUT, Papers by PROUT (*Ann. Phil.*, 6 (1815) 321) and others, Alembic Club Reprint No. 20, Livingstone Ltd., Edinburgh, 1932.
3 T. THOMSON, *An Attempt to Establish First Principles of Chemistry by Experiment*, 1825.
4 J. R. PARTINGTON, *A History of Chemistry*, Vol. IV, Macmillan and Co., London, 1964, p. 226.
5 J. W. DÖBEREINER, *Ann. Phys. Chem.*, 15 (1829) 301.
6 M. VON PETTENKOFER, see Ref. 4, p. 884.
7 J. W. VAN SPRONSEN, *The Periodic System of Chemical Elements: A History of the First Hundred Years*, Elsevier Publishing Co., Amsterdam, 1969, p. 338.
8 J. A. R. NEWLANDS, *Chem. News*, 12 (1865) 83.

9 J. A. R. NEWLANDS, *On the Discovery of the Periodic Law and on the Relations among the Atomic Weights*, Spon, London, 1884.
10 W. ODLING, *Quart. J. Sci.*, 1 (1864) 642.
11 W. A. SMEATON, *J. Roy. Inst. Chem.*, 88 (1964) 271.
12 L. MEYER, *Ann. Chem. Pharm.*, Suppl. VII (1870) 354. Meyer's first periodic table, drawn up in July 1868, was intended for a new edition of his book *Moderne Theorien der Chemie*, but this table was not published.
13 D. I. MENDELEEFF, *J. Russ. Chem. Soc.*, 1C (1869) 60.
14 D. I. MENDELEEFF, *J. Prakt. Chem.*, 106 (1869) 251.
15 D. I. MENDELEEFF, *Z. Chem.*, 12 (1869) 405.
16 O. N. PISARZHEVSKY, *Mendeleeff, His Life and Work*, Foreign Language Publishing House, Moscow, 1954, pp. 73, 91.
17 N. A. FIGUROVSKY, *D. I. Mendeleeff*, Izv. Akad. Nauk S.S.S.R., Moscow, 1961, pp. 44–51.
18 A. E. BÉGUYER DE CHANCOURTOIS, *Compt. Rend.*, 54 (1862) 757.
19 C. A. WINKLER, *J. Prakt. Chem.*, 34 (1886) 177.
20 D. I. MENDELEEFF, *Principles of Chemistry*, London, 1905, p. 496.

CHAPTER 9

*Consequences of solving the problem of atomic weights (II).
Birth of the structure theory*

Like the discovery of the Periodic Law, the growth of the theory of the structure of molecules was a gradual process extending over the greater part of the 19th century. The process began long before agreement on the question of atomic weights was reached but was greatly accelerated after that event.

Chemical structure theory has been described as one of the greatest achievements of the human mind—the greatest non-mathematical theory of science—comparable in stature to Newton's great synthesis of classical mechanics. It differed from the latter in that it was not the achievement of one man but the result of the efforts of many, though a few stood out pre-eminently as the principal architects of the structure theory. By the study of chemical reactions—the transformation of one substance into another—and with the atomic and valency theory as guiding principles, chemists were eventually able to discover the manner in which atoms are arranged in molecules even though the atoms, being not more than about 10^{-8} cm. in diameter, are far beyond the reach of vision.

Thus from the tiny seed represented by the simple but fundamental atomic notions of Higgins and Dalton sprang a mighty tree. Structure theory has been and continues to be one of the most successful and productive of all theories of science. In this brief sketch, designed primarily to show it as one of the consequences of fixing atomic weights, it will not be possible to do more than give in barest outline the story of its beginning.

Higgins and Dalton introduced the notion of atom-to-atom linking and represented simple molecules like sulphur trioxide and carbon dioxide by graphical formulae in which symbols were placed in juxtaposition.

Dalton assumed that the arrangements of several like atoms about a central atom of another kind were determined by the mutual repul-

References pp. 154-155

sion of the former atoms. On this basis he inferred a linear structure for carbon dioxide which happens to be the true structure of this molecule. For three like atoms arranged about a fourth he assumed a planar triangular structure. There was no experimental evidence to support the idea that the atoms were so arranged. Wollaston's brilliant and in many ways prophetic attempts to understand the three dimensional arrangements of atoms in solids were like Dalton's, purely speculative and made too soon to enter the stream of chemistry.

Higgins' formula for sulphur trioxide

Dalton's formula for carbon dioxide

Up to about 1820 it was generally believed by chemists that each substance possessed a unique chemical composition which could be represented by a unique formula. This belief was upset when in 1823 Gay-Lussac[1] and Liebig[2] discovered that silver fulminate had the same composition as silver cyanate which had been described and analysed a year earlier by Wöhler[3]. Thinking there might have been some mistake, each repeated his analyses only to confirm them. While Berzelius at first chose to reserve his opinion on the discovery, Gay-Lussac interpreted it straightforwardly as meaning that the same kinds and numbers of atoms must be arranged differently in the two substances. It was reasonable to expect that this would result in substances of different properties. In one respect the two substances could have hardly been more unlike. On being heated, silver cyanate melts; the same treatment causes silver fulminate to explode violently. Long afterwards the difference in the atomic arrangements in these two substances was shown to be a very simple one indicated by the two formulae AgONC (silver fulminate) and AgCNO (silver cyanate).

While continuing his work on the salts of cyanic acid (HCNO) Wöhler[4,5] discovered that ammonium cyanate could, by the simple process of heating, be transformed to urea—a compound of identical composition. So astonished was Wöhler at the formation of this compound which had previously been supposed to depend in some unknown way on "vital" energy supplied by living organisms, that he postponed publishing his discovery for several years. He, like Gay-

Lussac, interpreted the difference between the two compounds as arising from different arrangements of the same numbers and kinds of atoms. After Gay-Lussac had discovered that racemic and tartaric acid are also identical in composition, Berzelius became convinced of the reality of the phenomenon of isomerism. In the last instance, the difference between the two compounds is a more subtle one: an aqueous solution of one compound (tartaric acid) causes a rotation of the plane of polarisation of polarised light, that is, it is optically active, whereas racemic acid produces no such effect. The origin of this difference, then unknown to either Gay-Lussac or Berzelius, will be described in a later section.

For the different pairs of substances, identical in composition but different in properties, Berzelius proposed the name isomers* a term used ever since. These discoveries made a deep impression on many chemists. Although the nature of the different arrangements of the atoms in the various pairs of isomeric compounds had not been revealed, the discovery of their existence may be said to mark the beginning of the structure theory.

A second vital step towards the building up of the structure theory was the discovery of radicals—groups of atoms that retain their identity throughout a series of chemical transformations. The first instance of a clearly identified radical was probably the cyanogen radical (CN) discovered by Gay-Lussac[6]. He found that on heating mercuric cyanide it yielded an inflammable gas composed of carbon and nitrogen $(CN)_2$ which he called cyanogen. He identified the presence of the CN group in hydrogen cyanide (HCN), cyanogen chloride (ClCN), cyanogen iodide (ICN) and the metal cyanides (KCN).

A more complex radical, that became known as the benzoyl radical, was discovered by Wöhler and Liebig[7] in 1832 when studying the oil of

---

* Berzelius distinguished between isomerism and polymerism—where the relative numbers of the atoms are the same but the absolute numbers are different. Faraday discovered that a gas, first noticed by Dalton, and now known as butylene, had the same composition as ethylene. He noted "this is the first time that two gaseous compounds have been supposed to exist differing from each other in nothing but density". One is twice the density of the other. Berzelius explained the difference as arising from the fact that the molecule of butylene (C_4H_8) contained twice as many atoms as ethylene (C_2H_4). It was in fact only where the two substances could be vaporised without decomposition that the distinction between isomerism and polymerism could be made. Other methods for determining molecular weight had not then been devised.

bitter almonds (benzaldehyde). By the use of appropriate chemical reagents they were able to transform benzaldehyde into a whole series of closely related compounds in each of which a group, of atoms $C_{14}H_{10}O_2$ was present*. They decided to call the group, which we now write C_6H_5CO, the benzoyl radical. Using the present day formulae we may indicate the transformations carried out by Liebig and Wöhler as follows:

Oil of bitter almonds (benzaldehyde) C_6H_5COH was oxidised to benzoic acid C_6H_5COOH; treatment of benzoic acid with chlorine yielded benzoyl chloride C_6H_5COCl and this with ammonia yielded benzoic amide $C_6H_5CONH_2$. Throughout all these transformations the group C_6H_5CO remained intact; in the words of Liebig and Wöhler the group or radical "comports itself like an element". This phenomenon, sometimes referred to as the "persistence of structure", is a most important one. Were it not for its existence, the task of elucidating the structure of molecules would have been infinitely more difficult than it proved to be. Another radical discovered soon afterwards was the ethyl radical C_2H_5 present in ethyl chloride (C_2H_5Cl) and ethyl alcohol (C_2H_5OH). Chemists now began a wide-ranging search for other radicals hoping to find that, like the elements, they would prove to be limited in number. In this they were mistaken; the number of radicals seemed to be legion and the problem soon arose of how to find order and relationships among the radicals.

The theory of types was the result of attempts to classify organic compounds in terms of radicals. The classification was based on four fundamental or prototype molecules—ammonia, water, hydrogen and hydrogen chloride. It was the work of a number of chemists including that of A. W. Hofmann[8, 9]. From his studies of the reactions of ammonia with ethyl chloride, he reached the conclusion that the compounds obtained were the result of successively substituting the ethyl radical for a hydrogen atom in ammonia. These compounds he classified as the ammonia type.

---

* The discovery of isomers and radicals was made despite the fact that there was some uncertainty about atomic weights. As long as a set of atomic weights was uniformly and consistently applied to all the compounds being investigated, the relationships involved in the two discoveries were not obscured.

$$\left.\begin{array}{l}\text{H}\\\text{H}\\\text{H}\end{array}\right\}\text{N} \qquad \left.\begin{array}{l}\text{C}_2\text{H}_5\\\text{H}\\\text{H}\end{array}\right\}\text{N} \qquad \left.\begin{array}{l}\text{C}_2\text{H}_5\\\text{C}_2\text{H}_5\\\text{H}\end{array}\right\}\text{N} \qquad \left.\begin{array}{l}\text{C}_2\text{H}_5\\\text{C}_2\text{H}_5\\\text{C}_2\text{H}_5\end{array}\right\}\text{N}$$
ammonia ethylamine diethylamine triethylamine

The water type, the next to be formulated, was proposed by Williamson[10] who prepared a series of compounds, of which diethyl ether is an example, by treating ethyl iodide with the potassium salt of ethyl alcohol (C_2H_5OK). He formulated the compounds thus:

$$\left.\begin{array}{l}\text{H}\\\text{H}\end{array}\right\}\text{O} \qquad \left.\begin{array}{l}\text{C}_2\text{H}_5\\\text{H}\end{array}\right\}\text{O} \qquad \left.\begin{array}{l}\text{C}_2\text{H}_5\\\text{C}_2\text{H}_5\end{array}\right\}\text{O}$$
water ethyl alcohol diethyl ether

Gerhardt generalised the type theory by introducing two additional types, namely the hydrogen type which included the paraffin hydrocarbons and the hydrogen chloride type which included ethyl chloride:

$$\left.\begin{array}{l}\text{H}\\\text{H}\end{array}\right\} \qquad \left.\begin{array}{l}\text{C}_2\text{H}_5\\\text{H}\end{array}\right\} \qquad \left.\begin{array}{l}\text{C}_2\text{H}_5\\\text{C}_2\text{H}_5\end{array}\right\}$$
hydrogen ethane butane

$$\left.\begin{array}{l}\text{H}\\\text{Cl}\end{array}\right\} \qquad \left.\begin{array}{l}\text{C}_2\text{H}_5\\\text{Cl}\end{array}\right\}$$

These formulae might appear to have structural implications. The chemists of the time did not so regard them. Gerhardt who took a prominent part in formulating types was very sceptical about the possibility of ever being able to discover the structure of molecules. The theory of types was essentially classificatory.

Before any progress could be made in discovering how atoms were linked together within radicals and in fact how the radicals themselves were arranged in molecules, rules describing the capacity of one atom to link with others had to be discovered. In other words the concept now known as valency had to be developed. This depended on a knowledge of the true relative weights of atoms.

Though the proponents of the radical theory did not seem to doubt the reality of these entities, no one had ever isolated a radical as a stable substance. This does not imply that they did not try to prepare them.

References pp. 154-155

In fact in 1852 Frankland[11] (1825–1899) thought he had done so—that by treating zinc with ethyl iodide* he had prepared the ethyl radical (C_2H_5). Actually what he had prepared, in this mistaken belief, was butane (C_4H_{10}). In the series of experiments of which this was part, he achieved something of far greater significance. During the above experiment Frankland isolated a white crystalline compound ethyl zinc iodide ($Zn\ C_2H_5I$). When this compound is heated, a vapour is formed which condenses to a colorless, transparent liquid with a peculiarly penetrating odour. The substance, zinc diethyl, was the forerunner of a large class of compounds known as organo-metallics, many of which are volatile.

Frankland went on to prepare similar compounds of arsenic, antimony and bismuth. The origin of the concept of valency is to be found in this work. He drew attention to the relation between the formulae in the first two columns shows below:

		Modern as:
$As \begin{cases} O \\ O \\ O \end{cases}$	$As \begin{cases} C_2H_3 \\ C_2H_3 \\ C_2H_3 \end{cases}$	$(CH_3)_2O$
$As \begin{cases} O \\ O \\ O \\ O \\ O \end{cases}$	$As \begin{cases} C_2H_3 \\ C_2H_3 \\ O \\ O \\ O \end{cases}$	$(CH_3)_2O.OH$
$Sb \begin{cases} O \\ O \\ O \end{cases}$	$Sb \begin{cases} C_4H_5 \\ C_4H_5 \\ C_4H_5 \end{cases}$	$Sb(C_2H_5)_3$
$Sb \begin{cases} O \\ O \\ O \\ O \\ O \end{cases}$	$Sb \begin{cases} C_4H_5 \\ C_4H_5 \\ C_4H_5 \\ O \\ O \end{cases}$	$Sb(C_2H_5)(OH)_2$
ZnO	$Zn\ C_2H_3$	$Zn(CH_3)_2$
SnO	$Sn\ C_4H_5$	$Sn(C_2H_5)_2$

* $Zn + 2C_2H_5I \rightarrow Zn\ I_2 + \text{"ethyl"}$.

In writing these formulae Frankland used the Gmelin system of atomic weights, some of which were equivalents and others atomic weights.

$$H = 1 \quad C = 6^* \quad N = 14 \quad O = 8^* \quad Cl = 35$$
$$P = 31 \quad As = 75 \quad Sb = 122 \quad Zn = 32.5^* \quad Sn = 59^*$$

It was, for chemistry, a remarkable piece of good luck that, as one writer recently put it "the inconsistencies of the Gmelin system of 'atomic weights' should so often result in constitutional formulae displaying at a glance a valency principle".

Commenting on the above formulae, Frankland wrote: "When the formulae of inorganic compounds are considered, even a superficial observer is struck with the general symmetry of their construction; the compounds of nitrogen, phosphorus, arsenic and antimony especially exhibit the tendency of these elements to form compounds containing 3 or 5 equivalents of other elements and it is in these proportions that their affinities are best satisfied. In the ternal group we have:

NO_3, NH_3, PO_3, PH_3, PCl_3, SbO_3, SbH_3, $SbCl_3$, AsO_3, AsH_3, $AsCl_3$ and in the 5 atom group:

NO_5, NH_4O, NH_4I, PO_5, PH_4I.

Without offering any hypothesis regarding the cause of this symmetrical grouping of atoms, it is sufficiently evident from the examples given, that such a tendency or law prevails, and that no matter what the character of the uniting atoms may be, the *combining power of the attracting elements*, if I may be allowed the term, is always satisfied by the same number of these atoms"[11].

To describe the number of atoms which would combine with a single atom of arsenic for example, that is to express the idea of combining power, Frankland's contemporaries introduced the term *atomicity*. This term was widely used for a number of years to denote what we now call valency. The atomicity of arsenic was said to be three or five. The first use in print of the term valency, according to the O.E.D., was in an essay by an anonymous author "*Sigma*" in the *English Mechanic*, Vol. X, Nov. 19, 1869. However there appears to be some doubt about this. Historians of chemistry usually attribute the first use to Wichelhaus[12] (1868) who suggested the term valence (valenz) as an abbreviation for quantivalence.

* These are half the modern values of the atomic weights.

References pp. 154-155

Before the concept of valency could be made precise, that is before valencies could be unequivocally assigned to the atoms of different elements, it was essential that atomic weights should be known with certainty. In 1859, Odling[13] wrote: "All chemists are agreed that the molecule of methane contains four atoms of hydrogen but they disagree whether it contains two atoms of carbon having each the value of 6 or one atom of carbon only having the value of 12".

Once the atomic weight of an element was known with certainty it became a simple matter to decide its valence or valency. When the atomic weight of an element is divided by its equivalent weight (defined as that weight of the element which combines with 8 g of oxygen), the result approximates to a small whole number—1, 2, 3, *etc*. This is the valency of the element. Just as an element may, in its different compounds, have different equivalent weights, so it may have different valencies. For example the valency of arsenic in some of its compounds is five; in others it is three. The same is true of phosphorus.

A definition of valency may be arrived at by another approach—through an examination of the compounds of hydrogen. A general molecular formula for the binary compounds of hydrogen may be written AH_n. For compounds where $n = 1$, no compound of the type $A_m H$ is known where m is greater than 1. In other words an atom of hydrogen is never linked directly to more than one atom of any other element and is said to have a valency of one or to be monovalent. All elements that form hydrides of the type AH are also monovalent. In compounds of the type AH_n n is the valency of A: thus the valency of oxygen is two (OH_2); nitrogen three (NH_3) and carbon four (CH_4).

The discovery that the valency of carbon is four, like so many other scientific discoveries, was made independently by several individuals. Perhaps the most notable among them was Kekulé (1829–1896) one of the great architects of the structure theory. He explained his ideas in a paper entitled *"On the Constitution and Metamorphoses of Chemical Compounds and on the Chemical Nature of Carbon"*[14] in which he wrote: "If we look at the simplest compounds of this element CH_4, CH_3Cl, CCl_4, $CHCl_3$, $COCl_2$, CO_2, CS_2, CNH, we are struck by the fact that the quantity of carbon which is considered by chemists as the smallest amount capable of existence—the atom—always binds four atoms of a monatomic, two of a diatomic element, so that the sum of the chemical units of the elements combined with one atom of carbon is always equal to four. We are thus led to the opinion that carbon is tetra-atomic or tetrabasic*.

Kekulé

"In the case of substances containing several carbon atoms we must assume that at least some of the atoms (of other elements present) are held bound by the affinities of the carbon atoms and that *the latter are themselves linked together* whereby part of the affinity of the one (carbon atom) is necessarily tied by an equally large part of the affinity of the other". The decisive step here was recognition of atom-to-atom linking of carbon. The carbon chain was to prove an almost universal feature of the molecular structure of organic compounds. A simple example of a compound with a three carbon chain is propyl alcohol Kekulé's graphical formula for which is shown in Fig. 7.

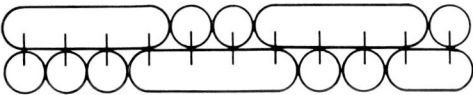

Fig. 7. Kekulé's formula for propyl alcohol (1861).

* The terms diatomic, triatomic, used in reference to atomicity (valence) are not to be confused with their earlier use in discussing Avogadro's work. The confusing term tetrabasic for tetravalent was fortunately abandoned.

References pp. 154-155

It will be seen that Kekulé employed a rather inconvenient method for representing the structure of the molecule. Each atom is represented by a symbol whose length is proportional to its valency. These symbols, placed together in the manner shown, can hardly be said to convey a very clear picture of the structure of the molecule. Indeed it is doubtful whether, at first, he really intended it to. Nevertheless the structural feature that the formula was designed to illustrate was of profound importance.

Kekulé's account[15], given long afterwards, of how he arrived at the idea of chains of carbon atoms is an interesting example of the origin of a scientific discovery:

"During my stay in London I resided for a considerable time in Clapham Road in the neighborhood of Clapham Common. I frequently however, spent my evenings with my friend Hugo Muller at Islington at the opposite end of the metropolis. We talked of many things, but most often of our beloved chemistry. One fine summer evening I was returning by the last bus, 'outside', as usual, through the deserted streets of the city, which are at other times so full of life. I fell into a reverie (traumerei), and lo, the atoms were gamboling before my eyes. Whenever, hitherto, these diminutive beings had appeared to me, they had always been in motion; but up to that time I had never been able to discern the nature of their motion. Now, however, I saw how, frequently, two smaller atoms united to form a pair; how a larger one embraced the two smaller ones; how still larger ones kept hold of three or even four of the smaller; whilst the whole kept whirling in a giddy dance. I saw how the larger ones formed a chain, dragging the smaller ones after them but only at the ends of the chain. I saw what our past master, Kopp, my highly honored teacher and friend, has depicted with such charm in his "*Molekularwelt*"; but I saw it long before him. The cry of the conductor, 'Clapham Road', awakened me from my dreaming; but I spent a part of the night in putting on paper at least sketches of these dream forms. This was the origin of the '*Structure Theory*'."

Valency theory became a powerful tool in the working out of the details of molecular structure. A very brief and necessarily incomplete description of one example must suffice. For some time Kekulé was greatly puzzled about the molecular structure of benzene, a colorless, volatile liquid discovered by Faraday in 1825. Chemical analysis revealed that its empirical formula, which merely shows the ratio of the numbers of different atoms present in the molecule, is CH, in seeming contradiction to the idea that the valency of carbon is four. The molecular weight calculated from vapour density measurements is 78. The molecular formula which shows the total number of atoms in the molecule is therefore C_6H_6. Kekulé found it impossible to reconcile this formula with any structure involving a terminating chain of carbon atoms. He then had the brilliant idea that the chain of atoms was a closed one, taking the form of a hexagonal six carbon ring with a hydrogen atom attached to each carbon. His first graphical formula for benzene is shown in Fig. 8(a).

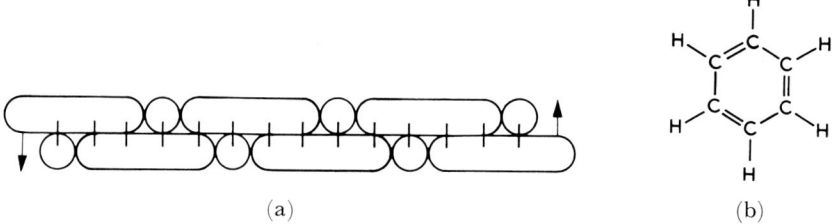

Fig. 8. (a) Kekulé's first graphical formula for benzene (1865) and (b) a later one for comparison.

Like the graphical formula for propyl alcohol it does not convey structural information as clearly as his later version shown for comparison. The ring structure for benzene proved correct and of profound importance in understanding the molecular structure of an extremely large class of substances which included dyes, drugs, explosives and natural products.

Kekulé was not alone in discovering that the carbon atom is tetravalent and able to form chains. This discovery was made independently and almost simultaneously by Couper (1831–1892), a wealthy Scotsman and wandering scholar, but little known as a chemist, who was undoubtedly the first to represent these ideas clearly by means of graphical formulae of the kind in use today. Fig. 9(a) shows his first attempt to represent the structure of propyl alcohol[16].

Shortly afterwards, reviving a practice begun by Higgins in 1789, he substituted solid lines for dotted lines to represent chemical bonds. This formula (Fig. 9(b)) is very close to the modern graphical formula,

References pp. 154-155

departing therefrom in the fact that Couper[17] used one dash to represent bonds to two and three hydrogen atoms. Another difference from the formula now accepted is that O–OH appears instead of OH—as a result of assuming that the atomic weight of oxygen is 8. Fig. 9(c) shows the modern formula for comparison.

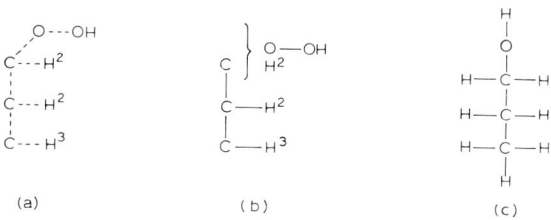

Fig. 9. Couper's formulae for propyl alcohol (1858) and a modern formula for comparison.

Couper, through an accident of fate, was anticipated in the publication of his ideas of chemical structure by Kekulé. At the time, both were working in Paris in the laboratory of Wurtz. Couper had handed Wurtz a note intended for publication in a journal (*Compt. Rend. Acad. Sci.*). For reasons that are still only partly clear, Wurtz delayed sending in the communication with the result that Kekulé's paper which appeared in May 1858 had priority. Part of the delay is believed to have been due to the fact that Wurtz did not approve of Couper's ideas. Certain it was that the delay led to a serious quarrel between the two and to Couper's dismissal from the laboratory. Couper returned to Scotland where he suffered a serious mental illness said to have been aggravated by his worries over the loss of priority.

In point of fact, it is difficult to disentangle the contributions of three individuals to the theory of chemical structure. The Russian chemist Butlerov (the first to use the term "*chemical structure*"), Kekulé and Couper all worked in Wurtz's laboratory during the year 1858. Butlerov expounded his ideas at a meeting in Paris on the 17th February 1858, but it is not known whether Kekulé or Couper was present on that occasion or not.

Several months later, Couper explained his chemical theory[17] at a meeting of the Chemical Society of France held on 23rd June 1858[19]. In the meantime, Kekulé's classical paper had appeared in the May issue of *Liebig's Annalen*[14]. It seems unlikely that there was no exchange of ideas between the three men. It has been suggested that it might be fairest to refer to the Couper–Kekulé–Butlerov theory of chemical structure.

There is, however, no doubt about Couper's claim to priority in the use of graphic formulae to represent the structure of the molecules of carbon compounds. His method of representing molecular structure was undoubtedly superior to that of Kekulé. Kekulé clung to his "sausage" formulae for a number of years after Couper's formulae were published but he was not as obstinate as Dalton had been over the use of symbols and eventually adopted the more modern system. The first textbook[18] to make general use of graphic formulae was Butlerov's *Introduction to the Study of Organic Chemistry*.

Butlerov laid great emphasis on the revolutionary idea—obvious to us now but not to all chemists then—that for any given structural formula there existed only one compound and that for any individual compound only one structural formula could be written.

Several minor variations of Couper's original graphic formulae were adopted before present day usage was fully established. For example, A. Crum Brown who at one time studied under Kekulé, used symbols reminiscent of those used by Dalton in the fully expanded graphic formulae (Fig. 10) for oxalic acid, published in the *Transactions of the Royal Society of Edinburgh* in 1864. In spite of their clarity, it would seem that Crum Brown's formulae did not mean to him what they mean to us today. He wrote of them: "I shall here explain the graphic notation which I imply to represent constitutional formulae and by which it is scarcely necessary to remark I do not mean to indicate the physical but chemical position of the atom".

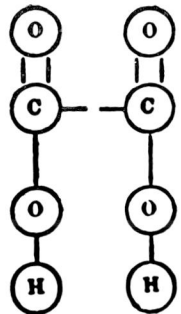

Fig. 10. Oxalic acid (Crum Brown, 1864).

References pp. 154-155

In their groping towards an understanding of the structure of molecules, chemists at this time seem to have been extremely reluctant to think of molecules as being built of atoms arranged in space in a manner that could be discovered. This is certainly the impression gained from reading their comments on graphical formulae.

Writing about this time, Frankland observed: "It must be carefully borne in mind that these graphic formulae are intended to represent neither the shape nor the supposed relative positions of the constituent hypothetical atoms". Kekulé himself in 1859, when referring to what he called rational formulae, wrote: "They do not express the constitution of the body; they are only expressions for the metamorphoses of a body and a comparison of different bodies, and are in nowise intended as an expression of the constitutions or the arrangements of the atoms in the actual substance". In other words, the formulae were regarded as reaction rather than structural formulae. This being so, Kekulé and others who held his views wrote different formulae for the one compound to indicate the different kinds of reactions or transformations it could be made to undergo. Not all of the nineteen formulae for acetic acid mentioned earlier originated from the adoption of different atomic weights for different atoms. Some of them were reaction formulae.

Much of the reluctance to accept structural formulae as representing real entities was philosophical in character and reminiscent of the situation of the atomic theory itself. In the "atomic debates" the argument centered on whether atoms were real or convenient hypothetical units that had no existence outside the human mind. A precisely similar question arose in regard to molecules: do structural formulae represent the arrangements of atoms in molecules or are they nothing more than conventional representations of what is experimentally known about substances, *e.g.* their reactions. In other words, is chemical structure something discovered or invented? Is it something that exists in nature independently of us or is it only an idea in our minds?

It is not easy to decide exactly at what period chemists generally began to believe in the reality of the physical structure of molecules in which the atoms were linked together atom to atom in the manner represented by graphical formulae. A case may be made for giving Couper the credit for being the first to do this in so far as his formulae very closely resemble the modern ones. His tragedy was that mental illness prevented him from developing his ideas. As far as explicit expression of belief is concerned Williamson's statement is probably the first expression of the modern view. Writing in 1852 he commented: "Formulae may be used as an actual image of the constituent atoms in a compound, as an orrery is an image of what we conclude to be our planetary system". As related in an earlier chapter, Williamson was an ardent supporter of the atomic theory and the leading protagonist of the theory in the 1869 debate of the London Chemical Society.

Some writers[20], mainly Russian, claim the honor of originating the structure theory for Butlerov. (See also ref. 21.) Butlerov first used the term "*chemical structure*" when he read a paper entitled "*The Chemical Structure of Natural Substances*" at the Conference of German Natural Scientists at Speyer in 1861. He defined it as follows: "Proceeding from the fact that each chemical atom bears a definite amount of chemical force [valency] with which it participates in forming a body [molecule], I would call that chemical binding, or the manner of the mutual uniting of atoms in a complex body [molecule], the chemical structure. In thus defining chemical structure, Butlerov integrated the concepts of atom, valency and interatomic binding. He emphasised the idea that each compound possesses a unique molecular structure and that the chemical properties of a compound depend on its molecular structure. It has been said that Butlerov was the first to suggest that it should be possible to derive a structural formulae for a substance by a study of the different methods of synthesising it. In an article on isomerism he explained that though the atoms in isomeric molecules are the same in number and kind, they are influenced by neighboring atoms, depending on chemical structure to acquire differing "chemical significance, *i.e.* differing behaviour in chemical reactions".

To return to the main question, there is little doubt that in 1865 A.W. Hofmann had a clear mental picture of the structure of some molecules. In a lecture before the Royal Institution that year, he constructed models with croquet balls of different colours to represent different atoms. A diagram of one of his models is shown in Fig. 11. He does not seem to have doubted that the molecules he demonstrated possessed planar structures, that is, the centers of all the atoms lay in one plane. We now take up the question whether the atoms are in fact so arranged.

References pp. 154-155

The existence of optical activity, that is the ability of a substance to rotate to the left or right of the plane of polarisation of plane polarised light, was first clearly recognised in crystals of quartz by the French

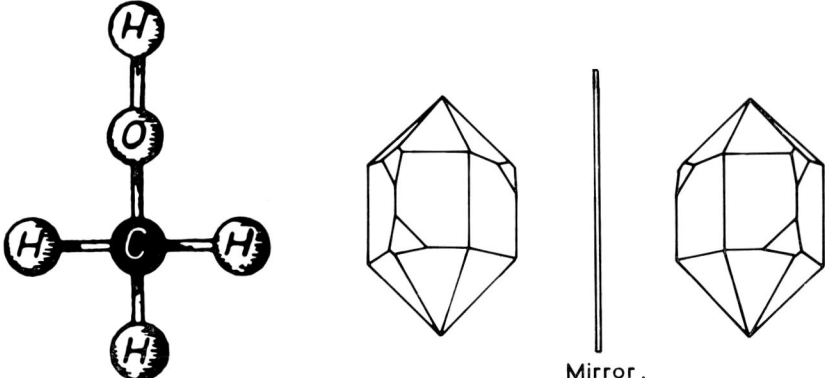

Fig. 11. Methyl alcohol (ball and stick model, Hofmann, 1864).

Fig. 12. Right- and left-hand quartz crystals.

physicist Biot. He found that some crystals of quartz rotated the plane of polarisation to the left, others to the right. Hauy had some years previously shown that quartz crystals exist in two forms distinguishable by the disposition of certain small facets and related to one another as an object and mirror image. Sir John Herschel related these two observations by showing that one form of quartz crystal rotated the plane of polarisation to the left, while the mirror image crystal rotated it to the right.

Biot also discovered that various liquids such as turpentine and solutions (*e.g.* of sucrose and tartaric acid) also possess the property of rotating the plane of polarisation in one sense or the other. Since solutions of these substances in optically inactive solvents exhibit optical activity, it was concluded that the phenomenon must originate in the molecules of the solute. Gay-Lussac and Mitscherlich had shown that tartaric and racemic acid are isomeric; a solution of the former is optically active while that of the latter is not.

All these facts were known at the time Pasteur (1822–1895) began his studies of tartaric acid. Pasteur noticed something that, surprisingly enough, Mitscherlich, an eminent crystallographer, had overlooked— namely, that crystals of tartaric acid, like quartz, existed in mirror image forms (Fig. 13). Working with a salt of tartaric acid, sodium

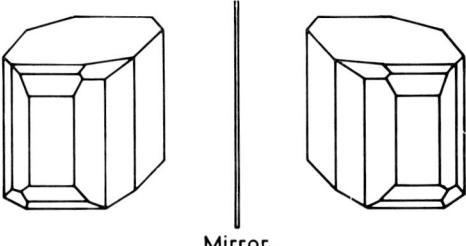

Fig. 13. Mirror image forms of crystals of tartaric acid.

ammonium racemate, Pasteur found that this too, when allowed to evaporate, slowly formed two kinds of crystals. Here is his account of what he observed:

"I carefully separated the crystals which were hemihedral to the right from those which were hemihedral to the left and examined their solutions in the polarising apparatus. I then saw with no less surprise than pleasure that the crystals hemihedral to the right deviated the plane of polarisation to the right and those hemihedral to the left deviated to the left; and when I took an equal weight of each of the two kinds of crystals, the mixed solution was indifferent towards the light in consequence of the neutralisation of the two equal and opposite deviations". Summarising his research on this subject in 1860, Pasteur[22] wrote (see also ref. 23):

"Are the atoms in dextra-rotatory tartaric acid grouped according to the spiral of a dextrose helix, or placed at the summits of a regular tetrahedron? We are unable to reply to these questions. But what cannot be doubted is that there is a grouping of atoms according to an order of dissymmetry to a non-superposable image". In other words, the molecules must exist in forms related to one another as an object to its non-superposable mirror image, that is, in dyssymmetric forms.

The idea that four atoms combined with a fifth were located at the corners of a tetrahedron was not new. It seems to have been first suggested by Wollaston in the Bakerian Lecture to the Royal Society in 1812 when he predicted that one day chemists would have to think

References pp. 154-155

of atomic arrangements in three dimensions. The tetrahedral arrangement of four hydrogen atoms about carbon in methane was suggested by Kekulé but he did not bring forward any experimental evidence to support the idea.

Butlerov also proposed that the four valence bonds of carbon are directed towards the vertices of a tetrahedron and used the idea to explain what he thought was an isomeric pair of $C_2H_5.H$ and $CH_3.CH_3$ (a notion now known to be illusory).

The answer to the question posed by Pasteur was found independently and almost simultaneously by van 't Hoff[24] (1852–1911) and le Bel[25] (1847–1938) in the form of a proposal that when four atoms are attached to a fifth, the four are located at the corners of a regular tetrahedron. Van 't Hoff, using the customary graphical formulae, discussed the number of isomers possible for molecules of the following types:

(1) CH_3R'
(2) $CH_2R'R''$
(3) $CHR'R''R'''$ Where R', R'', R''' represent different groups of atoms such as CH_3, OH, COOH, *etc.*

If the groups are all in the same plane as the central carbon atom, the structures to be expected would be those shown in Fig. 14:

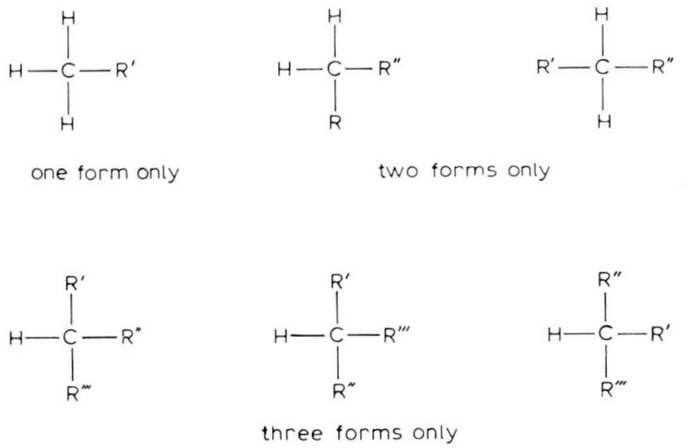

Fig. 14. Planar arrangements of four groups about a carbon atom.

On the other hand, if the four groups are placed at the corners of a regular tetrahedron about C, the number of forms for (2) and (3) are one and two, respectively (Fig. 15).

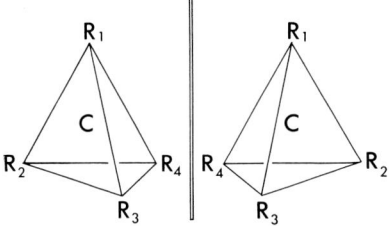

Fig. 15. Tetrahedral arrangement of four groups about a carbon atom (mirror image forms).

Experiments with compounds of the type $(CR_1R_2R_3R_4)$ e.g. lactic acid, show that there are two forms only, but more importantly, they are each optically active but in opposite senses:

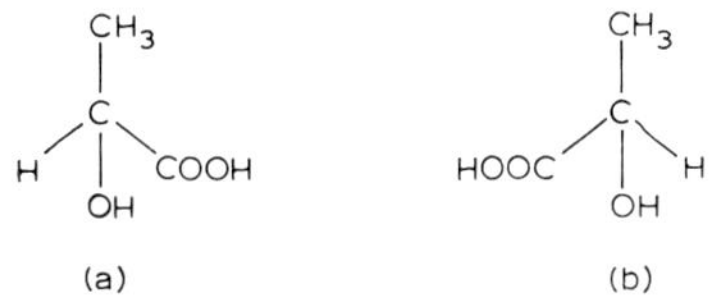

$$\begin{array}{cc} \text{(a)} & \text{(b)} \end{array}$$

Fig. 16. Mirror image forms of lactic acid.

(b) is a reflection of (a), i.e. a mirror image of (a) and the two are not superposable. Each of these forms is found in nature: the dextro-rotatory form in muscle, the levo-rotatory form in the fermentation of sucrose by *Bacillus acidi*. When lactic acid is synthesised in the laboratory from inactive substances, a mixture of the two forms (a racemate) results. These can be separated by a chemical method devised by Pasteur. A carbon atom to which four different groups are attached, is described as an asymmetric carbon atom and the presence of one such atom in a molecule means that it is possible for the molecule to exist in non-superposable image, and therefore optically active, forms. Although the graphic formulae in Fig. 14 represent planar molecules they also may be used without change to represent tetrahedral molecules if they are thought of as the projection of a tetrahedral structure on to the plane of the paper. This is the present day convention.

References pp. 154-155

The discovery of the existence of optically active isomeric molecules and the explanation of their structure in terms of molecules related to one another as non-superposable mirror images probably did more than any other single discovery to convince chemists of the reality of atoms arranged in three dimensions to form molecules.

At every stage in the growth of the atomic molecular theory there were doubters and sceptics and this stage was no exception. Kolbe[26, 27], a distinguished chemist, was perhaps one of the most vociferous opponents of the theory of van 't Hoff and le Bel. "A Dr. J. H. van 't Hoff of the Veterinary School at Utrecht has no liking it seems, for exact chemical investigation. He has considered it convenient to mount Pegasus (apparently borrowed from the Veterinary School) and to proclaim... when he is on the chemical Mt. Parnassus which he reached by bold flight, how the atoms appear to him to be arranged in space... it is not possible to criticise this work even half way thoroughly because the play of fantasy in it dispenses completely and entirely with factual basis and is absolutely unintelligible to the sober scientist". How wrong he was! Suffice it to say in conclusion that underlying principles and the results of chemical structure theory have now been fully confirmed by the completely independent and equally powerful method of structural analysis based on the diffraction of X-rays by crystals. This is another story—beyond the scope of this book.

REFERENCES

1 J. L. Gay-Lussac, *Ann. Chim.*, 27 (1824) 196.
2 J. Liebig, *Ann. Chim.*, 33 (1826) 207.
3 F. Wöhler, *Ann Phys.*, 1 (1824) 117.
4 F. Wöhler, *Ann. Phys.*, 12 (1828) 253.
5 F. Wöhler, *J. Sci. Arts*, 25 (1828) 491.
6 J. L. Gay-Lussac, *Ann. Phil.*, 7 (1816) 350.
*7 F. Wöhler and J. Liebig, *Ann. Pharm.*, 3 (1832) 249. English translation, *Am. J. Sci. Arts*, 26 (1834) 261.
8 A. W. Hofmann, *Phil. Trans.*, 140 (1850) 93.
9 A. W. Hofmann, *J. Chem. Soc.*, 11 (1859) 252.
10 A. W. Williamson, *J. Chem. Soc.*, 4 (1852) 106.
*11 E. Frankland, *Phil. Trans.*, 142 (1852) 417.
12 C. H. Wichelhaus, *Ann, Chim.*, Suppl. 6 (1868) 259.
13 W. Odling, *J. Chem. Soc.*, 11 (1859) 107.
*14 A. Kekule, *Ann.*, 106 (1858) 129.

15 A. Kekule, *Ber.*, 23 (1890) 1302.
*16 A. S. Couper, *Phil. Mag.*, 16 (1858) 114.
17 A. S. Couper, *Ann. Chim. Phys.*, Sér. 3, 153 (1858) 484.
18 A. M. Butlerov, *Introduction to the Study of Organic Chemistry*, Kazan, 1864. German translation, Leipzig, 1868.
19 J. Jacques, *Bull. Soc. Chim. France*, Sér. 5, 20 (1953) 528.
20 L. Gumilevskii, *Alexander Mikhailovich Butlerov*, Moscow, 1952, p. 152.
21 H. M. Leicester, *J. Chem. Educ.*, 36 (1959) 328.
22 L. Pasteur, *Leçons de Chimie Professées en 1860*, Soc. Chim. de Paris, 1861.
23 J. R. Partington, *A History of Chemistry*, Vol. 4, Macmillan and Co., London, 1964, p. 752.
*24 J. H. van 't Hoff, *Arch. Néerl. Sci. Exact. Nat.*, 9 (1874) 445.
*25 J. A. Le Bel, *Bull. Soc. Chim. France*, Sér. 2, 22 (1874) 337.
26 H. Kolbe, *J. Prakt. Chem.*, 15 (1877) 473.
27 H. Kolbe, *J. Prakt. Chem.*, 24 (1881) 405.

* The papers marked with an asterisk have been reprinted in O. T. Benfey (Ed.), *Classics in the Theory of Chemical Combination*, Dover Publications Inc., New York, 1963.

CHAPTER 10

Epilogue

In the foregoing account I have attempted to sketch briefly the development of the atomic theory up to about the year 1870 when, to those chemists who accepted it, the theory may have seemed to have taken a form that would require little further change. Of course, nothing could have been further from the truth. The closing years of the 19th century witnessed discoveries that were to revolutionise our views about the structure of matter. Over the last 100 years great advances have taken place in our understanding of the nature and structure of atoms. Many of these advances have been described by Schonland[1] whose account is, in some ways, complementary to this one, though it does not go beyond 1933.

My original intention was to describe the development of the atomic theory up to the time when the existence of atoms was more or less universally taken for granted. However it became evident that it would not be easy to decide when this happened but it was clearly much later than 1870. Even after the turn of the century there were a number of outstanding scientists —men like Berthelot, Mach and Ostwald who, despite the great advance made by Cannizzaro, remained sceptical of the atomic theory, believing that atoms were "mental artifices" or figments of the imagination. Of the three, Ostwald* (1853–1932) was perhaps the most notable. In his "*Faraday Lecture*" to the Chemical Society in 1904 Ostwald[2] made a case for chemistry without the atomic theory. "It is possible" he said "to deduce from the principles of chemical dynamics all the stoichiometrical laws; the law of constant proportions, the law of multiple proportions and the law of combining weights. You all know that up to the present

---

* For his work on catalysis and for his investigations into the fundamental principles governing chemical equilibria and rates of reactions, Ostwald was awarded the Nobel Prize in 1909.

time it has only been possible to deduce these laws by the help of the atomic hypothesis. Chemical dynamics has, therefore, made the atomic hypothesis unnecessary . . . .".

What finally brought the sceptics around to a belief in the existence of atoms? There is room for difference of opinion about the answer to this question but most of those who have attempted to answer it would agree that a number of discoveries in physics had much to do with it.

For Ostwald it was, on his own admission, an understanding of the nature of the Brownian movement and the results of counting atoms and molecules that caused him to change his mind about the atomic theory. The Brownian movement was first discovered by microscopic observation of pollen grains suspended in water. (See Chapter 1, p. 9). It is exhibited by particles whose diameters lie in the region of 5×10^{-5} cm and it is immaterial whether they are suspended in a gaseous or liquid medium*. The entirely random and incessant movement of the particles has been described as "an approximate image of the movements of gas molecules postulated by the kinetic theory of gases"[3].

It arises from fluctuations in the bombardment of the particles by the molecules of the surrounding liquid or gaseous medium. At a given instant of time, the total momentum of molecules striking a particle in one direction is greater than that of molecules striking the particle in the opposite direction and hence the particle moves a short distance, depending partly on its mass, in the former direction. Then the same kind of thing happens all over again but in a different direction.

On the assumption that the mean kinetic energy of a particle suspended in a fluid medium is the same as the mean kinetic energy of a molecule in the same medium at the same temperature, Einstein[4,5], in 1905, derived an expression for the displacement of a particle Δx in a time interval τ which lent itself to the determination of what came to be known as Avogadro's number (N), the number of molecules (or atoms) in 1 gram molecule (or gram atom) of a substance.

Shortly afterwards (1908), Perrin[6,7], using a uniform suspension of gamboge particles made a number of determinations of N of moderate accuracy. In two series of experiments he obtained values of 6.82×10^{21} and 6.88×10^{23}, respectively.

* The movement is easily observed in a suspension of carbon particles (diluted India ink) or cigarette smoke in air.

References p. 163

These were not the first determinations of this important number. As early as 1865 J. Loschmidt, on the basis of the kinetic theory of gases and some assumptions about the packing of molecules in liquefied gases, made a crude estimate of the number of molecules in 1 cm³ of a gas at standard temperature and pressure—a number that came to be known as Loschmidt's number L*.

The interest of Perrin's results lies not so much in the novelty of the method he used to obtain them as in the fact that they agreed so well with results obtained by entirely different methods. By 1909, Avogadro's number had been measured by at least eight different methods** with results that agreed remarkably well and left no doubt that something real was being counted. N clearly represented a fundamental constant of nature of the greatest importance.

Mainly as a result of these achievements Ostwald[9] was moved to write in the preface of a new edition (1909) of his *Outlines of General Chemistry*, a frank admission of his conversion: "I am now convinced", he said, "that we have recently become possessed of experimental evidence of the discrete or grained nature of matter for which the atomic hypothesis sought in vain for hundreds and thousands of years. The isolation and counting of gaseous ions on the one hand which have crowned with success the long and brilliant researches of J. J. Thomson [1856–1940] and on the other, the agreement of the Brownian movement with the kinetic hypothesis established by many investigators and most conclusively by Perrin, justify the most cautious scientist in now speaking of the experimental proof of the atomic theory of matter. The atomic hypothesis is thus raised to the position of a well founded scientific theory and can claim a place in a textbook intended for use as an introduction to the present state of our knowledge of General Chemistry".

Once one knew the number of atoms in one gram atom of an element it was then possible from a knowledge of the density of the element

---

* This is converted to Avogadro's number by multiplying by 22,412 (the volume in cm³ occupied by one gram molecule of a gas at standard temperature and pressure). On the Continent the term Loschmidt's number and the symbol L were later used for Avogadro's number.

** Since 1909 the accuracy of measurements of N has been continually improved. The generally accepted value of N at the present time is 6.02252×10^{23}. Virgo[8] in 1933 noted that there had been eighty different determinations of N by twenty different methods.

in the solid or liquid state, to form an approximate idea of the size of an atom of that element. A simple calculation using diamond as an example shows that the diameter of an atom of carbon has somewhere between 1×10^{-8} and 2×10^{-8} cm which puts the atom forever beyond the range of even the most powerful optical microscopes.

Despite their small size, it became possible, in the first few years of this century, to see the effects produced by single atoms.

In 1903 Crookes devised a very simple instrument known as a spinthariscope. It consisted of a small slip of glass coated with a layer of finely crystalline zinc sulphide mixed with an extremely small quantity of radium or other radioactive material which acted as a source of α particles. When the layer of zinc sulphide is viewed with a high powered lens in a darkened room by an observer whose eyes have been dark adapted, he sees a remarkable display of intermittent tiny star-like flashes of light or scintillations against a black background. (See also ref. 11.) Quantitative studies of these scintillations by Regener[12] and Rutherford (1871–1937) and Geiger[13] established the fact that each flash arises from the violent impact of a single α particle on the surface of a zinc sulphide crystal. An α particle is an atom of helium that has lost its two electrons. As each radium atom disintegrates, it ejects an α particle endowed with great energy because of its high velocity and it is because of its great energy that it produces effects that are visible. (See also ref. 8.)

Rutherford and Geiger also detected and counted individual particles by means of the ionization they produced in a gas at low pressure—an effect upon which the Geiger counter is based. They found that this method and the scintillation method of counting α particles gave approximately the same result. Counting the number of α particles from a known mass of radium formed the basis of an approximate method for determining Avogadro's number. Rutherford and his students also made great use of the scintillation method in exploring the nucleus of the atom.

Another important device for observing effects produced by single atoms, invented about this time, was the Wilson[14] cloud chamber. It was based on the fact that α particles in their passage through air leave behind a trail of ionized molecules. If the air through which the α particles pass, is saturated with water vapor and then allowed to cool and become supersaturated by suddenly expanding it, droplets of water condense on the ionized molecules in the wake of the α particles to produce an easily visible trail several centimeters in length depending on the energy of the α particle. This technique also found great use in the study of the structure of the atom.

The last example of a discovery that may have greatly helped to confirm a belief in the existence of atoms depends not on effects produced by single atoms but on cooperative effects produced by large numbers of atoms arranged in a regular manner.

As mentioned earlier, the idea that crystals are ordered assemblies of particles of some kind goes back to Kepler, Hooke, Huygens and Wollaston; the particles were assumed to be spherical, or occasionally, ellipsoidal but were not characterised further. They were just particles—particles of alum, calcite or boracite or whatever substance was under discussion.

Probably the first person to attribute the form and symmetry of crystals to explicit and stated arrangements of chemical atoms was Barlow[15] who suggested for example that in sodium chloride each atom of sodium was surrounded by six equidistant octahedrally arranged chlorine atoms. Similarly each chlorine atom was surrounded by six sodium atoms as shown in Fig. 17.

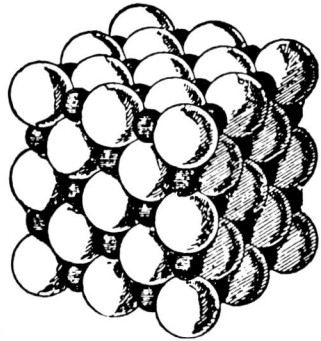

Fig. 17. Barlow's suggested structure for sodium chloride (1897).

This was a remarkably good guess but at the time (1897) there was no way of testing experimentally the proposed structure.

The opportunity came in 1912 with von Laue's[16] (1879–1960) epoch making discovery of the diffraction of X-rays by crystals. Sommerfeld had shown that if x-rays were electromagnetic waves, their wave lengths would be in the region of 1×10^{-8} cm—about the same order as the dimensions of atoms.

Aware of the speculations that in crystals atoms were arranged in a regular manner, von Laue had the brilliant idea that a crystal might serve as a diffraction grating for X-rays. At his suggestion two of his research assistants Friedrich and Kipping carried out an experiment in which a sharply defined beam of X-rays was allowed to pass through a thin slice of crystal. As von Laue had anticipated, they were able to record on a photographic plate placed at a small distance from the crystal, a symmetrical pattern of spots formed where the diffracted rays struck the plate*. With this relatively simple experimental arrangement, two facts of the greatest importance were established: 1, that X-rays are electromagnetic waves of very short wavelength; 2, that atoms are regularly arranged in crystals.

Within about a year of von Laue's discovery, W. L. Bragg[17] and his son W. H. Bragg demonstrated that X-rays could be used to discover the manner in which atoms are arranged in crystals. At first they chose for study simple crystals, sodium chloride and diamond. They showed that the atoms in sodium chloride are arranged as Barlow had predicted and for the first time the distance between atomic centers was measured. The arrangement of carbon atoms in diamond proved to be also a relatively simple one: every carbon atom is surrounded by four others arranged at the corners of a regular tetrahedron, thus confirming the ideas of le Bel and van't Hoff about the stereochemistry of carbon. It would be out of place to attempt to describe the technique of X-ray crystal analysis here. Suffice it to say that X-ray crystal analysis has not only substantiated the results of the structure theory of the organic chemists to a remarkable degree but has revolutionised our understanding of the arrangements of atoms in minerals and those inorganic compounds to which the structure theory is not applicable.

By about 1914, the year of the Braggs' discoveries, evidence for the existence of atoms was overwhelming; so much so that when, some 40 years later Müller[18] produced direct images of iridium atoms (Fig. 18), in the finely pointed tip of an iridium wire by means of his field ion microscope, it was probably the feat itself rather than what was revealed that stimulated the greatest interest.

* The first crystal used was hydrated copper sulphate but it proved unsatisfactory and was substituted by the more symmetrical crystal of zinc sulphide.

References p. 163

Each dot in Fig. 18 is the image of an atom in the surface layer of a metal crystal. The formation of these images, representing as it does a magnification of more than a million, is one of the most remarkable achievements of physics in recent years.

Fig. 18. Ion micrograph of a crystal of iridium (Erwin Müller, *J. Appl. Phys.*, 1965).

Even in the purest of metal crystals, some of the lattice points are occupied by foreign or impurity atoms but it is not possible by visual inspection of the micrograph to discover the chemical identity of any particular atom. However, Müller[19] has shown that by combining a field ion microscope with a time-of-flight mass spectrometer capable of single particle sensitivity, it is possible to achieve the ultimate goal of microanalysis—the chemical identification of a single atom selected by the observer.

REFERENCES

1 B. SCHONLAND, *The Atomists 1805–1933*, Clarendon Press, Oxford, 1968.
2 W. OSTWALD, Faraday Lecture: Elements and Compounds, *J. Chem. Soc.*, 85 (1904) 506.
3 L. J. BODAZEWSKI, *Dingler's Polytech. J.*, 239 (1881) 325.
4 A. EINSTEIN, *Ann. Phys.*, 17 (1905) 549.
5 A. EINSTEIN, *Ann. Phys.*, 19 (1906) 289, 371.
6 J. PERRIN, *Compt. Rend.*, 146 (1908) 967.
7 J. PERRIN, *Compt. Rend.*, 147 (1908) 475, 530, 594.
8 S. E. VIRGO, *Science Progress*, 27 (1933) 634.
9 W. OSTWALD, *Outlines of General Chemistry*, translated by W. W. TAYLOR, Macmillan and Co., London, 1912.
10 W. CROOKES, *Proc. Roy. Soc.*, 71 (1903) 405.
11 J. Elster and W. GEITEL, *Phys.* [2], 4 (1903) 439.
12 E. REGENER, *Verhandl. Deut. Phys. Ges.*, 10 (1908) 78.
13 E. RUTHERFORD AND H. GEIGER, *Proc. Roy. Soc.*, 81 (1908) 141.
14 C. T. R. WILSON, *Proc. Roy. Soc.*, 85 (1911) 285.
15 W. BARLOW, A Mechanical Cause of the Homogeneity of Structure and Symmetry, *Proc. Roy. Dublin Soc.*, 8 (1897) 527.
16 M. T. F. VON LAUF, *Sitz. ber. Akad. München*, 1912, p. 303.
17 W. L. BRAGG, The Structure of Some Crystals Indicated by their Diffraction Patterns, *Proc. Roy. Soc.*, 89 (1914) 248.
18 E. W. MÜLLER, S. NAKAMURA, O. NISHIKAWA AND S. B. McLANE, *J. Appl. Phys.*, 36 (1965) 2496.
19 E. W. MÜLLER AND T. T. TSONG, *Field Ion Microscopy*, Principles and Applications, American Elsevier Publishing Co., New York, 1969, pp. 7, 131 and 293.

Name Index

Adet, P. A. 44
Amontons, G. 65
Ampere, A. M. 73, 78, 79, 83, 101, 106
Anaximander 3
Antropoff, A. von 132
Archimedes 30
Aristotle 3
Austin, W. 52
Avogadro, A. 62, 65, 71, 72, 73, 74, 75, 76, 77, 78, 79, 80, 81, 82, 83, 87, 90, 91, 97, 99, 101, 102, 105, 106, 107, 108, 109, 116, 143, 157, 158, 159

Bacon, F. 14, 29, 43
Bailey, C. 13
Barlow, W. 160, 161, 163
Benfey, O. T. 155
Bergman, T. 22
Bernouilli, D. 20, 21, 25, 26
Berthelot, D. 156
Berthollet, A. B. 42, 49, 50, 56, 62, 66, 78, 79
Berzelius, J. J. 62, 75, 80, 81, 82, 87, 88, 89, 90, 91, 92, 93, 94, 96, 100, 101, 102, 104, 105, 120, 121, 122, 136, 137
Biot, J. B. 150, 151
Black, J. 31, 34, 35, 36, 43, 67
Bodazewski, L. J. 163
Boscovitch, R. 45
Boyle, R. 14, 16, 17, 18, 20, 22, 26, 30, 31, 32, 36, 38, 43, 50, 65, 81
Bracciolioni, P. 14
Bragg, W. H. 161, 163
Bragg, W. L. 161
Brodie, B. C. 115, 117
Brown, R. 9
Brown Crum, A. 147
Butlerov, A. 147, 149, 152, 155

Cannizzaro, S. 107, 108, 109, 111, 112, 113, 114, 116, 117, 122, 127, 156
Cardwell, D. S. L. 26, 63
Cavendish, H. 36, 37, 40, 55, 67
Chancourtois, A. E. B. de 123, 128, 131, 134
Charles, J. A. C. 65
Charleton, W. 14, 22, 26
Cicero 6
Clausius, R. J. 116, 117
Comte, I. A. M. 100, 103
Couper, A. S. 145, 146, 147, 148, 155
Crookes, W. 159, 163
Crossland, M. P. 83, 102

Dalton, J. 21, 40, 45, 46, 47, 48, 49, 50, 51, 52, 53, 54, 55, 56, 57, 58, 59, 60, 61, 62, 63, 64, 65, 69, 70, 71, 72, 75, 77, 78, 81, 83, 84, 87, 89, 90, 102, 109, 112, 120, 135, 136, 137, 147
Davy, H. 37, 48, 49, 50, 61, 62, 68, 69, 81, 102
Delametherie 80
Democritus 1, 14
Dobereiner, F. J. W. 121, 122, 133
Dulong, P. L. 93, 94, 95, 96, 97, 98, 99, 102, 103, 104, 105, 113, 114
Dumas, J. B. 98, 99, 100, 102, 103, 105, 107, 109, 122, 127

Einstein, A. 157, 163
Elster, J. 163
Epicurus 1, 14, 15

Faraday, M. 115, 137, 144
Figurovsky, N. A. 134
Forbes, R. J. 43
Foster, G. C. 124
Fourcroy, A. F. 42

Fox, R. 103
Frankland, E. 115, 117, 140, 141, 148, 154
Friederick, W. 161

Garnett, T. 48
Gassendi, P. 14, 15, 16
Gaudin, M. A. 100, 101, 103, 106, 107
Gay-Lussac, J. L. 65, 66, 67, 68, 69, 71, 74, 75, 78, 80, 83, 88, 102, 107, 116, 136, 137, 151, 154
Geiger, H. 159, 163
Geitel, W. 163
Geoffroy, E. G. 22
Gerhardt, C. 105, 106, 107, 117, 139
Gibbs, F. W. 43
Gladstone, J. H. 122
Glauber, J. R. 29
Gmelin, L. 102, 141
Gough, J. 47
Greenaway, F. 63
Gregory, J. C. 75, 83
Grotthus, C. J. D. 59, 64
Guericke, O. von 16
Guerlac, H. 4, 29, 30, 43
Gumilevskii, L. 155

Hales, S. 36
Hall, A. R. 31, 43
Halley, E. 23, 26
Hariot, T. 14
Hartley, H. 63, 102, 117
Haüy, R. J. 25, 150
Helmont, J. B. van 34
Henry, W. 49, 50
Heraclitus 3
Herschel, J. 150
Higgins, B. 45
Higgins, W. 44, 45, 46, 61, 62, 63, 64, 67, 135, 145
Hinrichs, G. O. 123
Hoff, J. H. van 't 112, 117, 152, 154, 155, 161
Hofmann, A. W. 138, 149, 150, 154
Hooke, R. 16, 20,24, 25, 26, 85, 160
Humbold, A. von 67, 68
Huygens 25, 160

Jacques, J. 155
Jorpes, J. E. 102

Kargon, R. H. 26
Kekulé, A. 104, 106, 112, 114, 115, 117, 142, 143, 144, 145, 146, 147, 148, 152, 154, 155
Kent, A. 43
Kepler, J. 23, 24, 25, 26, 160
Kipping 161
Klaproth, M. H. 55
Knight, D. M. 114, 117
Kolbe, C. 154, 155
Kopp, E. 82, 113, 117, 144
Kuhn, T. S. 43, 64
Kundt, A. A. 113, 117

Laplace, P. S. 66
Latham, R. 8
Laue, M. von 160, 161, 163
Laurent, A. 105, 106, 117
Lavoisier, A. L. 27, 34, 37, 38, 39, 40, 41, 42, 43, 44, 52, 55, 61, 67, 80, 118
Le Bcl, J. A. 154, 155, 161
Leicester, H. T. 155
Lemay, P. 26, 103
Lemery, L. 18
Leucippus 1
Liebig, J. von 102, 103, 136, 137, 138, 154
Lomonosov, M. 21, 26
Loschmidt, J. 158
Lucretius, C. 1, 14, 15, 21, 22

McKie, D. 43
McLane, S. B. 163
Mach, E. 156
Macquer, P. J. 30, 43
Margraff, A. S. 31
Marignac, J. C. G. 109
Maxwell, J. C. 116, 117
Mayow, J. 36
Meldrum, A. N. 46, 55, 63
Mendeleef, D. I. 117, 123, 126, 127, 128, 129, 130, 131, 132, 134
Menschutkin, B. N. 26, 127
Meyer, L. 107, 117, 123, 126, 127, 131, 134
Meyer, E. von 107
Millikan, R. A. 111
Mills, E. J. 115, 116, 117
Milt, C. de 117
Mitscherlich, A. E. 92, 93, 100, 102, 151

Monge, G. 68
Morrow, S. I. 102
Morveau, Guyton de 42
Müller, E. W. 161, 162, 163
Muller, Hugo 144

Nakamura, S. 163
Nash, L. 51, 63
Newlands, J. A. R. 117, 123, 124, 126, 128, 131, 133, 134
Newton, I. 14, 15, 18, 19, 20, 21, 22, 23, 26, 33, 34, 50, 81, 96, 135
Nishikawa, O. 163
Nobel, A. 156

Odling, W. 115, 117, 119, 123, 124, 125, 133, 134, 142, 154
Oesper, R. E. 103
Ostwald, W. 156, 157, 158, 163
Owen, R. 48

Parmenides 3, 5
Partington, J. R. 13, 46, 53, 62, 63, 90, 101, 103, 120, 127, 133, 155
Pasteur, L. 151, 152, 153, 155
Pauling, L. 43, 81, 83, 116, 117
Pavesi, A. 107
Perrin, J. 157, 158, 163
Petit, A. T. 93, 94, 95, 96, 97, 98, 99, 103, 104, 105, 113, 114
Pettenkofer, M. 122, 133
Pisarzhevsky, O. N. 134
Priestley, J. 36, 37, 38, 39, 43
Proust, L. G. 56, 63
Prout, W. 120, 121, 133

Read, J. 43
Regener, E. 159, 163
Regnault, H. V. 96, 103
Richter, J. B. 55
Ritchie, A. D. 117
Roscoe, H. E. 46, 112, 117
Russell, B. 5, 9, 13
Rutherford, D. 36
Rutherford, E. 159, 163

Sambursky, S. 13
Sarton, G. 118
Scheele, C. W. 36, 37
Schneer, C. 26
Schonland, B. 163
Schorlemmer, C. 46
Seguin, A. 68
Seitz, E. 103
Smeaton, W. A. 134
Soddy, F. 62, 64
Sommerfeld, A. 160
Spronsen, J. W. van 133
Stahl, G. E. 33

Taylor, Sherwood 29, 43
Taylor, W. W.
Thackray, A. W. 26, 51, 56, 58, 63, 64
Thales, 2
Thenard, L. J. 68
Thomson, J. J. 158
Thomson, T. 48, 49, 50, 59, 60, 61, 63, 64, 84, 102, 120, 133
Torricelli 16
Trevelyan, R. C. 4
Tsong, T. T. 163

Vavilov, S. I. 6
Virgo, S. E. 158, 163

Waals, J. D. van der 21
Warburg, E. 113, 117
Watts, J. 66
Weltzein, K. 106
Wenzel, C. E. 55, 100
Wheeler, T. S. 62, 63
Whyte, L. L. 13
Wichelhaus, C. H. 141, 154
Williamson, A. W. 114, 115, 117, 139, 149, 154
Wilson, C. T. R. 159, 163
Winkler, C. A. 130, 134
Wohler, F. 136, 137, 138, 154
Wollaston, W. H. 63, 84, 85, 86, 102, 135, 152, 160
Wordsworth, W. 47
Wurz, A. 146

Subject Index

Académie des Sciences, 63
air, composition of 40, 67
   homogeneous mixture 49
   nature of - Greek view 2
   pump, invention of 16
alchemy, Newton's attitude to 33
   origin of some chemical techniques 29
   source of new substances 29
alum, crystals 24
   pure substance of antiquity 28
   solution 16
analytical chemistry, beginnings of 30
arsenic, alloy with copper 28
   vapor density of 99
assayers' contribution to chemistry, 31
atoms, attractive forces between 31
   fortuitous concourse of 6
   Greek views of 1
   indivisibility of 1, 15, 58, 112
   linking by hooks 10, 18
   linking by valency bonds 146
   relative weights of 46, 51, 52, 58, 59, 60, 78, 111
   size of, first attempts to estimate 23
"Atomic debates", 114, 148
atomic heats, 95, 113
   volumes 82
   weights, arithmetical relations between 120, 121, 122
   weights, first table of 52
atomicity (valence), 141, 143
atomism and atheism, 13, 15
Avogadro's hypothesis, 71, 72, 73, 74, 75, 76, 77, 78, 79, 80, 81, 82, 83
   number, methods for estimating 157, 158

Bakereian Lecture (Wollaston), 85, 152
benzene, discovery of 144
   structure of 145

Berthollide compounds, 56
boron trifluoride, combination with ammonia 68
Boyle's law, 18
British Association for the Advancement of Science, 63
Brownian movement, 9, 157

caloric theory, 17, 49, 72
carbon dioxide, absorption of, by water and lime water 35, 36
   in air 36
chemical analysis, qualitative 30
   quantitative 31
chemical compounds, qualitative composition of 29
   quantitative composition of 30, 31
Chemical Society and Newlands, 124
chemical structure, 146
chemistry, definition of 26
cinnabar, composition of 30
classification of the elements, 118, 119
cloud chamber, 159
combustion, phlogiston theory of 33
   role of air in 38
Comparative View of Phlogistic and Anti-Phlogistic Theories (Higgins), 44
compound, Proust's definition of 56
crystals, alum 24
   calcite 25
   cleavage 25
   diamond 161
   diffraction of X-rays by 161
   isomorphism 92
   snow 24
   sodium chloride 160, 161
   symmetry of 25
cupellation, 28

De Rerum Natura (Lucretius), 1, 4, 6, 14, 15

diffusion, Newton's explanation of 20
dissymmetry, 152
Dobereiner's triads, 121
dualistic theory, 79, 81, 82

elective attractions, 21
electrochemical series, 21
elements, classification of 118, 119
   definition of 41
   Greek concept of 2
   prediction of missing 130
   symbols for 44, 89
equivalent weights, 59, 87, 97

field-ion microscope, 161
"fixed air", 36
formulae, empirical 60
   graphical 146
   molecular 60, 145

gamboge, colloidal particles of 157
gases, discovery of common elementary 36
   kinetic theory of 116
   law of partial pressure 50
   laws of Boyle 64, Charles 64, Gay-Lussac 69
   solubility of 35, 36, 50, 51
   structure of, Dalton's view of 49
germanium, discovery and properties of 130
gold, specific physical and chemical properties of 32
graphical formulae, 145, 146

heat, caloric theory of 17
   kinetic theory of 17
   specific 93, 94, 95, 96, 97
hygroscopic salts, 18
hypothetico-inductive method, 54

iatrochemistry, 29
Introduction to the Study of Organic Chemistry (Butlerov), 147
isomerism, mirror-image 153
isomorphism, 92
isotopes, 130

Karlsruhe Conference, 106
kinetic theory of gases, 116

lactic acid, optical activity of 153
law of Dulong and Petit, 94
law of octaves, 124
laws of chemical combination, 53
lime, limestone, limewater 34, 35, 36
Loschmidt's number, 158

magnesia alba, 34
Manchester Literary and Philosophical Society, 48
matter, conservation of 42
   correlation of physical properties with structure 10, 11
   indestructibility of 3
   kinetic theory of 116
   unity of 2, 61, 133
mercury barometer, invention of 16
metals, solution of, in acids 18
Meteorological Observations and Essays (Dalton), 47
Micrographia (Hooke), 24
microscope, electron 23
   field-ion 161
   optical, Hooke's observations of crystals with 24
mirror-image isomerism, 153
missing elements, prediction of existence of 130, 131
mixtures, 56
molecular heats, 115
molecular models, 149
molecular volume, 82
molecular weight, 111
molecule, Avogadro's use of the term 76
   definition of 101, 112
molecules, three dimensional structure of 154
multiple proportions, law of 58

New System of Chemical Philosophy (Dalton), 56, 59, 62
nitrogen, discovery of 36
   oxides of 69
Nobel Prize for Chemistry, 156
nomenclature, chemical 42

octaves, law of 124
opal, origin of color of 23
optical activity, 150

Opticks or Treatise on Reflections etc. (Newton), 21
organometallic compounds, 140
Outlines of General Chemistry (Ostwald), 158
oxalic acid, composition of and formula for 60
oxygen, discovery of 36

particle, α 159
periodic law, discovery of 122, 123
persistence of structure, 138
philosopher's stone, 28
phlogiston theory, 33
phosphorus, vapour density of 99
plenum theory, 3
polymerism, 137
positivism, 100, 102
Principia (Newton), 42
Principles of Chemistry (Mendeleef), 127
Prout's hypothesis, 120

quantification of chemistry, 34

radicals, 137
rare gases of the atmosphere, 123
Royal Institution, 48
Royal Society, Davy medal to Newlands 124
    Royal medal to Dalton 63

scepticism of the atomic theory, 102, 114, 115
Science Museum, South Kensington, 38
snow crystals, 23
Société d'Arcueil, 66

Society of Friends, 47
solubility of gases, 35, 36, 50, 51
solutions, analogy with gases 112
specific heat, 93, 94, 95, 96, 97
specific properties, 32
spheres, close packing of 24, 25, 85
spinthariscope, 159
structure, persistence of 138
structure theory, 144
substance, definition of 27
sulphur, vapour density of 99, 100
sulphur dioxide, composition of 45, 46

tartaric acid, optical activity of 151
telluric helix, 123
tetrahedral configuration, 152
Traité Elémentaire de Chimie (Lavoisier), 42
transmutation, 32, 33
triads, Dobereiner's 121

unity of matter, 2, 61, 133

vacuum, 4, 16
valency, 141, 142
vapor density, Dumas' method of measurement 98
volume diagrams (Gaudin), 101
volume relationships of combining gases, 69

water, composition of 40, 52, 67, 74, 77
    Greek view of the nature of 2

X-ray crystal analysis, 154, 161